CLICK TO WIN!

Clicker Training for the Show Ring

CLICK TO WIN!

Clicker Training for the Show Ring

COLLECTED ARTICLES FROM THE *AKC GAZETTE*

By Karen Pryor

Foreword by George Berger
Publisher, The American Kennel Club

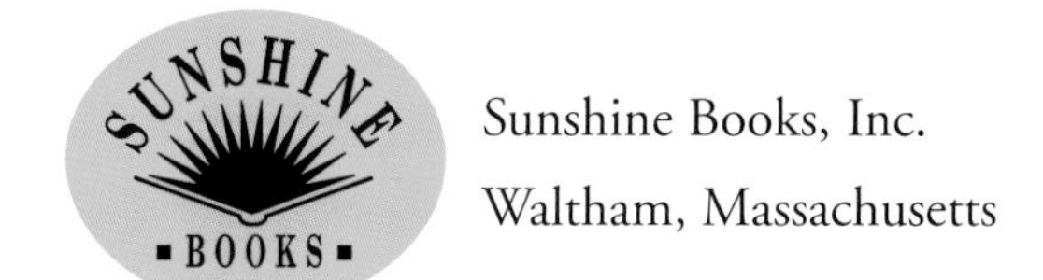

Sunshine Books, Inc.
Waltham, Massachusetts

With the exception of Chapters 7, 9, and 15, the articles in this book appeared in the *AKC Gazette* between 1996 and 2001. The appendices are based on similar material in Karen Pryor's *Don't Shoot the Dog! The new art of teaching and training, Revised Edition,* Bantam Books, 1998 or on other instructional materials by Karen Pryor and distributed by Sunshine Books, Inc.

Click to Win!

Book design by Codesign, Boston

Library of Congress Cataloging-in-Publication Data is available.
Library of Congress Card Number: 2002090065

ISBN 1-890948-10-1

Sunshine Books, Inc.
49 River Street, Suite #3
Waltham, MA 02453
781-398-0754
www.clickertraining.com

10 9 8 7 6 5 4 3 2 1

CONTENTS

CONTENTS *continued*

FOREWORD

For many years, my family and I have had the good fortune of traveling to the Abaco Islands in the Bahamas. It is a glorious part of the world—as close to paradise as you can imagine—and we pray that time passes quickly from the end of one trip to the start of next.

Life changes refreshingly little in these islands, but on our arrival several years ago, we heard about something new and exciting that was going on up at Baker's Bay, a 40-minute boat ride from where we were staying: a group of behavioral scientists had set up a dolphin training center and the public was welcome to come and watch them work. Not only that, there were times of the day when you could swim with the dolphins in their in-ocean netted enclosure. That's all we had to hear. We were up and on our way well before eight o'clock the next morning. It turned out to be one of the most electrifying days of our lives.

We knew that dolphins are among the most intelligent of animals, but until you swim side by side with one, you can't fully appreciate its gentle yet immense power, its sensitivity, its grace. Our youngsters, who had been timid at first, cried when they had to come out of the wondrous pool. But there were more thrills ahead.

The scientists now began a training session. I didn't quite understand the technique at the time, but the sequence was always the same: the trainer gave a verbal or hand signal; the dolphin responded; the trainer immediately tooted on a whistle and tossed the dolphin a treat (a fish or two). Some animals progressed more quickly than others, of course, but all of them were excellent "students." I was absolutely fascinated and moved by this beautifully effective, positive training method.

Karen Pryor, son Ted, and a spinner dolphin, Hawaii

After the swimming and the training "show" and a perfect picnic lunch, I wandered back to the training area to see what more I could learn. The young trainers were delighted to talk about the dolphins, the idyllic lifestyle that they themselves enjoyed living and working in the Bahamas, and their hopes for ever-broader applications for the method they were using. They explained that this kind of training was formally known as operant conditioning, that they were "shaping" the animals' behavior. And they told me that without question the greatest practitioner of the process my family and I had witnessed that morning was someone named Karen Pryor, who had done most of her original work in Hawaii. They spoke with an unmistakable reverence for Karen, and told me that two books she had written, *Lads Before the Wind* and *Don't Shoot the Dog!* were must-reads.

How right they were! Within a week, I had read both books and become completely absorbed with the Pryor approach to training. When we returned from our trip, I immediately began what she calls clicker training with our somewhat stubborn (but loveable) Labrador retriever, Ben. To my delight and amazement, Ben caught on almost immediately. After only a few lessons, a dog that had previously gone into a heeling position only when it struck his fancy was now marching along consistently and smartly, in lockstep with me. In the past, Ben had come when called sluggishly, sometimes almost resentfully, but after just a few days of this training he sped to me as though I had a T-bone in each hand. I had, in effect, helped shape a new and better Ben, and it was thanks to Karen Pryor.

A couple of years later I made a career change, moving from a major magazine company to join the Publishing Division of the American Kennel Club. The monthly *AKC Gazette*, the organization's flagship publication, is a mainstay in the world of purebred dog fanciers nationwide. It has been in continuous publication since 1889, and, as such, is the oldest sporting journal in this country. The *Gazette* survives—and, indeed, thrives—because of its strict adherence to presenting accurate information in a quality format. We don't compromise.

As the new man in charge, I was mightily impressed with the lineup of regular contributors to the magazine, an all-star array of photographers, illustrators and authors. And—certainly you've guessed it by now—at the top of the list was Karen Pryor who, delightfully for me, continues to write a regular column for us.

Karen's home base is Watertown, Massachusetts, but she travels a great deal, making appearances, leading seminars, and conducting workshops—"spreading the word of clicker training," as she puts it. I've now met with her a couple of times here in our AKC offices in New York. She is a small, attractive woman who brims with enthusiasm, especially when she ever so gently describes her work and her hopes for its future. She is truly one of a kind: a scientist-trainer-author-visionary. I am proud to play some small role in providing a forum for her.

We give readers of the *AKC Gazette* a new world of opportunities through Karen Pryor's writing. Now, with this collection, that world opens to you, too.

George Berger
Publisher, The American Kennel Club

INTRODUCTION

I love dog shows, whether I'm competing or not. I love the dogs, the commotion, the glamour. I love to watch the intensity of the judges. I love the egalitarianism: anyone might win, as I once discovered for myself.

Clicker training offers the owners and handlers an alternative: a totally force-free way to establish whatever behavior you want, from a great hands-off stack to gaiting brilliantly on a slack leash, to pricking up the ears on cue and gazing warmly at the judge.

I went to my first big dog show because of my young Weimaraner, Gus, whom I'd bought as a puppy through an ad in the newspaper. We were there for the obedience trial and I had signed up for conformation too, just for fun. Gus did well in obedience. In conformation he beat all the other Weimaraners and then took a first in the larger Sporting Group category. What a thrill!

Gus enjoyed the show, too—one of his assets in any kind of competition was a continuously wagging tail. But many dogs do not love dog shows. They may be calm—they know their jobs, they don't flinch from the judge, they take the commotion in stride—but it's not their idea of fun. One reason is obvious. Skilled or novice, professional or amateur, handlers often feel they have to make the show dog do its job, in various uncomfortable ways. For the dog, going into the breed ring or practicing to do so often involves being pinched, poked, yanked, dragged, popped with short jerks of the leash, hauled skyward by the neck, and forced into awkward positions for long periods of time. Even with the sop of a stream of treats, it takes a game dog to tolerate the process for long.

Clicker training offers the owners and handlers an alternative: a totally force-free way to establish whatever behavior you want, from a great hands-off stack to gaiting brilliantly on a slack leash, to pricking up the ears on cue and gazing warmly at the judge. Clicker training is fun for the dog, and for the handler too. It's much easier to learn than the physical skills of leash handling and correction-based training, and for teaching the skills that the dog needs in the conformation ring—posing, moving correctly and in a straight line, standing still to be examined—it's very quick. Clicker training is a way of communicating to your dog exactly what you want, and will pay for. Once the dog understands the behavior, you can turn the responsibility over to him. It will be the dog who wants to gait correctly, stack like a champion, or hold ears, head, and tail the way they look best.

Clicker training is a way of communicating to your dog exactly what you want, and will pay for.

This book is based primarily on a collection of articles that were published over a five-year period in the AKC Gazette, a monthly magazine published by the *American Kennel Club*. Its chapters, most addressing specific show ring applications, are drawn from my own experiences and observations and those of my students. This is not a complete step-by-step guide to everything you might possibly want to know about clicker training or the conformation ring, but if you are interested in either, it should be enough to give you some ideas and get you started. (For tips and an overview of the principles behind clicker training, see the appendices at the end of this book.)

So put away the leash or tie it to your belt—from now on that leash is for safety only, not for education. Plan on working in short, five-minute segments; use a really delicious treat at first, cheese or ham or chicken, not kibble. Pick some silly thing to start with—a paw movement or a head dip, or touching a target—rather than something serious. That way you and your dog can learn the game without being too worried about the outcome of your first clicker training sessions.

Once you and the dog have had some success with a trick or two, you can start working on those show ring behaviors you've admired in other dogs and would love to help your own dog acquire. Clicker training is easy to learn, in part because mistakes don't do any harm; the worst that can happen is that the dog gets a click and a treat for nothing. Thousands of people have taught themselves how to use this method—

and besides, your dog will help you! (For more information and support, see the "Resources" section of this book, or visit our Web site, www.clickertraining.com, for online answers to frequently asked questions, lists of clicker trainers in your area, books, videos, and links to other sources of clicker guidance.)

Will clicker training help you win? I've heard exhibitors grumble that another dog beat theirs just because of "showmanship," but if showmanship means a happy, confident, well-displayed dog and a happy, confident handler, I'm all for it. If you were a judge, given two dogs of roughly equal merit physically, one full of confidence and enthusiasm and the other just going through its paces, which one would you put first? Clicker training can give you a dog that loves to show; a dog that brings its very best self into the ring; a dog that knows what to do and does it with gusto. If you are looking for those ribbons and trophies, the clicker dog's winning ways can provide you with a great head start.

Clicker training can give you a dog that loves to show; a dog that brings its very best self into the ring; a dog that knows what to do and does it with gusto.

Win or lose, you'll have the pleasure of knowing that it's not just you who's having fun at the dog show. And even more important, you may find, as so many of us have, that the clicker "game" creates a mutual respect and a level of communication and understanding that will make your life with your dog richer and more interesting for both of you.

Karen Pryor
January, 2002

CHAPTER I

SHAPING FOR THE SHOW RING

How to "shape" your show dog into a winner with a click and a treat

"Shaping" turns show dogs into winners. My friend Barbara loves Great Danes and had always enjoyed showing her dogs, but her newest dog—a female named Heather—presented a problem that Barbara had never encountered before. In her first puppy class appearance at eight months of age, as the judge leaned over to touch the dog, Heather ran behind Barbara and wouldn't let the man near her. She was disqualified because of her seemingly poor temperament. In fact, Heather was terrified of strangers, and it looked as if her show career would be over before it had begun. Barbara approached me with this question: How could we change this timid dog's behavior?

"Shaping" is scientific slang for building a particular behavior by using a series of small steps to achieve it. Shaping allows you to create behavior from scratch, without physical control or corrections but rather by drawing on an animal's natural ability to learn.

Lately many dog trainers have begun applying this technique—called operant conditioning—to canine tasks and sports.

To shape behavior rapidly and effectively you must use a distinct signal, such as a touch or a noise, that marks the instant the right action occurs. After the signal the animal is given something it likes, such as praise, petting, toys, or food.

To shape behavior rapidly and effectively you must use a distinct signal, such as a touch or a noise, that marks the instant the right action occurs.

Although praise and food conveys to the animal that you're pleased, the marker signal is actually the more important part of the process because it tells the animal in the moment *exactly* what it was doing that earned it the treat. This information makes it both possible and likely the animal will "do the right thing" again. Dolphin trainers use a whistle as their marker signal, or "conditioned reinforcer." Dog trainers most often use a clicker that's a descendant of the children's toy also known as a cricket.

How could this help Barbara? We arranged to meet at a dog show where Barbara had brought Heather just to get her used to the many new sights and sounds. Heather was certainly pretty, and the sights and sounds didn't seem to bother her. She gazed around with the aplomb typical of Danes—until I reached out to pet her. Then she shied like a horse and backed away to the end of her lead. I had no interest in why Heather behaved this way; my aim was to see if we could get her to react in a more appropriate manner.

We found a quiet spot, beyond the crowds. I bought some sausage at the hot dog stand on the show grounds (it's always wise to start this process with something truly delicious). Heather ate the sausage slices, but only if Barbara fed them to her; she wouldn't take them from me. I gave Barbara a plastic clicker and showed her how to begin the shaping procedure. Click, then treat; click, then treat. Teach the dog to expect the treat when it hears the click. Then I had Barbara walk the dog around for a few minutes, clicking whenever Heather appeared to relax.

Barbara took the clicker home. The next day, as I had suggested, she took Heather to a nearby shopping center. Whenever someone toward them, Barbara clicked, then stopped the dog and gave her a treat. Soon Heather was walking calmly toward approaching strangers.

Often, of course, peopled wanted to pet such a lovely animal. On the third day Barbara began letting people touch Heather on the back, clicking if Heather stood still. Heather quickly learned to stand still on purpose. From her viewpoint, she had Barbara all figured out: if she accepted petting Barbara would click and give her a treat every time.

The next weekend Barbara entered Heather in the second show of the dog's young life. Heather trotted calmly beside Barbara and stood politely while the judge looked at her teeth and felt her legs. Heather won her class. Three weeks later, Heather won a Puppy Class and beat several adult female Great Danes in the broader category, earning her first championship points.

A Pleasant Process

What happened to Heather seemed like a miracle. It wasn't. The clicker "explained" to Heather that she would be 'paid' for letting herself be touched by strangers. She discovered for herself that the process was harmless, even pleasant. The last time I saw Heather (again, at a show) she had dived into a crowd of teenagers and was reveling in being scratched and petted by six at once.

Becky's standard schnauzer, Dash, had a different problem: she wouldn't keep her ears and tail up. Like Heather, Dash was a very nice-looking bitch, and Becky felt she had great potential, but a schnauzer with its ears flat and its tail tucked in is not an impressive sight. Dash had long since lost interest in squeaky toys, so you couldn't fool her into pricking up her ears.

Becky got a clicker and taught Dash that click means treat. She then spent five minutes every evening playing with Dash. Every time Dash's ears went up, Becky clicked. A truck went by, the schnauzer pricked her ears: click, treat. Dash started pricking her ears on purpose. Soon Becky could wait to click until Dash kept her ears up for two second—then three, then five; then while the dog was posed and then while she was moving. Before long, whenever Dash saw the clicker her ears went up and stayed there.

Becky shaped Dash's tail carriage in exactly the same way. At first she gave Dash a click for any tail movement, then for a tail horizontal to the ground, then for a tail a little higher. Dash wagged her tail at first; later she started trying to lift it on purpose. They worked in five- or ten-minute sessions, first at home, then in parks, then in busy places among other dogs and people.

Becky also taught Dash to "self-stack"—to assume on her own the preferred posture that shows off a dog's body to best advantage. First she clicked for the back feet, until Dash always stopped with her hind legs in the right alignment. Then Becky "added" the front feet, later the head position. Dash preferred stacking herself to being pushed and pulled about, and was soon holding her pose like a model—even while a judge felt her coat and opened her mouth. She had quickly learned that she could rely on Becky to "pay" her for the job. Now, six months after the first click, Dash shines with pride and confidence; and she and Becky have won Best of Breed or Best of Opposite Sex at all four shows they've been in. (See Chapter Five for more on self-stacking.)

Try It Yourself

Although operant conditioning is different from conventional training, it is not particularly difficult. In fact, pet owners with no traditional dog training experience are often better at it than seasoned professionals who've grown to depend on other methods.

Try it for yourself. You don't need a clicker: rattle the change in your pocket, clink a spoon on a glass or cluck with your tongue to mark the instant you see a behavior in your dog that you want to see repeated. (It's best to save your voice for praise and affection; to dogs, a brief, unusual sound is a much clearer behavior marker than a spoken word.) Try to shape a simple task such as spin, roll over, or shut the door. Don't worry about how the dog "feels"; concentrate on what behavior you want to elicit. And remember to have fun! Shaping is a great game for you and your dog to play together, and fun is a great motivator for learning.

CHAPTER 2

BEST FOOT FORWARD

Improving your dog's gait for the show ring with a click and a treat

In the show ring, the way a dog moves can be more important than its coat, head, teeth, or any other physical characteristic. Movement shows best at the trot because, unlike the walk or the gallop, the trot is a symmetrical, balanced gait. In this motion, any unsoundness—a bad hip, a weak shoulder—will show up at once. To the trained eye, little defects in movement, such as spraddling the hind feet or crossing the front feet, can reveal poor proportions or structural flaws.

These flaws may not be evident under the thick coat of a collie or a Samoyed, or when a dog is standing still, but they are characteristics that affect the dog's stamina, endurance, lifelong health, and ability to perform the work for which it was bred. And assuming they are not the result of accident or injury, these characteristics may also be passed down to any offspring the dog might have, which is why an assessment of a dog's physical attributes is important to breeders and for dogs.

Judges, therefore, want to see each dog trot past them, which allows for a good side view of the animal, and directly toward and away from their line of sight, so they can gauge how closely the dog approaches its breed's ideal.

Because you'll be in motion and only able to see your dog from one angle, making progress toward a good-looking gait with a clicker is more easily accomplished by two people: one to click, one to walk or run with dog and give the treat.

All too often, however, the behavior of both dog and handler make fair evaluations impossible. If your dog trots on a diagonal (known as "crabbing") or lugs on its lead as if pulling a milk wagon, the judge has no chance to see anything but poor movement. Down go your dog's marks, down go your chances to win. But these handicaps can be avoided and your dog's movement improved by clicker training.

Team Training the Gait

The essence of clicker training is to use the clicker (or any other distinctive sound) to mark the exact moment your dog does something right. After clicking, tell your dog with petting, praise, and treats that you are pleased. The crucial information, however—what the dog did to earn all that—is provided by the marker signal. (As noted in Chapter One, you can ring a bell, blow a whistle, or jingle the coins in your pocket, but a clicker is both distinctive and easy to use, and it's important to use an artificial sound as your marker signal; research indicates that it is much clearer to dogs than any spoken word.)

Because you'll be in motion and only able to see your dog from one angle, making progress toward a good-looking gait with a clicker is more easily accomplished by two people: one to click, one to walk or run with dog and give the treat. Find a partner—perhaps someone who is also preparing to show a dog—and schedule some practice time. As you gait your dog, have your partner click when your dog does what you want. Stop instantly on the click and give your dog a treat of cubed cheese or diced chicken. After that, resume gaiting, giving your dog another chance to earn a click and treat.

It's important not to tease or bait your dog with the treat. Waving food around defeats the purpose of the clicker. It makes your dog watch and follow your hands instead of looking straight ahead; it also makes your dog think about food rather than what it is supposed to be doing to earn the click. Keep the treat out of sight until after the click. Also, don't click at the end of a run. At first click at random points during the exercise run, or you'll soon have a dog that looks bored during the run and elated only at the turnaround point.

Next, decide what needs improvement. Remember, you want your dog to be near you, perhaps even a little in front of you, and moving straight. You can put your partner alongside you to work on positioning your dog, or at either end of the track to work on its straight-line movement. You can also trade positions with your helper, so you can watch and click your own dog and see how it's coming along. Finally, keep the sessions short. Don't push yourself or your dog to the point of boredom or fatigue.

Working with a partner will enable you to teach your dog in just a few five-minute sessions to move in a straight line, to keep its ears and tail up, to have a happy look on its face, to move on a slightly loose lead. The goal is to hand these responsibilities over to your dog. If you use the clicker to explain what you want, your dog will be thrilled to oblige.

A loose lead is important. If you hold the lead taut, your dog will almost certainly resist, pulling sideways or backwards. Even a little resistance throws the gait off completely. Nevertheless, stringing your dog upon the neck seems to be the fashion these days. I recently saw a "professional" handler, in a major show, gaiting an Australian terrier with such a tight lead that the dog's front feet were completely off the ground. What the judge thought I can't imagine, but the dog was visibly miserable.

Problems You Can Fix

You can also use team training to correct flaws in your dog's gaiting. By clicking at the right moment, you can tell your dog that you want it to trot, not pace, or that it should keep its front paws aligned with its shoulders, not flying wide. Here's an example of how it's done.

Jennifer is a young St. Bernard that, like many big dogs, tended to shamble along with her head down. Her owner wanted Jennifer to look proud and confident by carrying her head high. We put an observer in the center of our practice space while the owner trotted Jennifer back and forth in a straight line. First the observer clicked after they'd taken only a few steps, and Jennifer's owner instantly stopped and gave the dog a treat. Then the process was repeated at random intervals three more times.

The next time the owner started off, Jennifer's head went up in expectation. Click and treat. Now, on each try, the observer clicked only when the head went up. In five minutes Jennifer was carrying her head high. She looked like a different dog—a winner. Did she "feel" different? Who knows? She looked happy, which wins points in any judge's eye.

The next step for Jennifer's owner would be to spend five minutes a day gaiting her at a trot while teaching her that the phrase "show time" means "Carry your head high, and you'll get a click and treat." Then the owner could, if she wished, replace the click with the word "good" and Jennifer would be ready for the ring—where, happily, you may talk to your dog and give it treats whenever you like.

By the way, can you run in a straight line with your head up? Most people can't, without practice. If you're unable to run straight, you may trip on your dog. If you look down at your dog, you'll make it run crooked. To fix this, draw a straight line on the floor on pavement with chalk, or on grass with flour. Then you and your partner should click each other for running beside the line without stepping on or over it. Both you and the dog will look calmer and more confident in the ring.

CHAPTER 3

LIKE A SHIP IN FULL SAIL

Clicker train your dog to travel in an extended trot to improve its performance in the ring

The most beautiful way for a four-legged animal to travel is in what horse trainers call an extended trot. Instead of just jogging along, the animal reaches, taking longer than normal strides with each step. An extended trot is not a *faster* trot: the cadence may not increase in the slightest. What does increase is the distance covered by every step, and the extra strength used in achieving that distance.

A dog in an extend trot seems to move powerfully, purposefully, and gracefully, almost floating over the ground. These are the dogs that catch the public's eye as soon as they enter the ring. You can hear the comments: "How proud he is!" "What a gorgeous dog!" "Look, you can tell she *knows* she's beautiful."

We get the impression of confidence, even pride, because of the function of this enhanced movement. In nature, the extended trot is what biologists call a display behavior. Display behaviors signal the message "Look at me!" You can see an extended trot when a stallion patrols the fence separating him from other horses. You can see it when a mature male dog notices and heads for another dog in the distance. You can see it sometimes when dogs compete in play, one when one captures the ball from another and gleefully trots off, head high, tail waving, with the prize.

In the ring, people hope for that look. Some people spend many hours "gaiting" the dog, trotting it up and down, luring it with food, encouraging it with the voice, trying to tease the dog into "showing" itself. Many handlers simply haul the dog's head into the air with the leash and then pull it forcibly along at the speed they think most likely to produce a decent-looking trot. Some breeders tend to select and show rather dominant individuals—"alpha animals," as biologists put it—because they go into the ring innately eager to be, literally, the top dog. These individuals, male or female, may give you a flashy, extended trot spontaneously. Of course they can also give you very dominant offspring, way beyond the management skills of average dog owners.

Clicker Training the Extended Trot

There's an easy way to get beautiful show ring gaiting from any well-built dog, without relying on an overabundance of dominance. You teach it to give you an extended trot—on purpose and on cue.

First, you need a way to identify for the dog what movement you want: the clicker will do that. Second, you need to be able to tell when the dog is beginning to give you the right kind of movement. However, it's hard to judge what a dog's legs are doing when you're looking down on it from your end of the leash. You can use mirrors, but moving and watching at the same time is still difficult. As with basic gait training, described in Chapter Two, the easiest solution is to find a partner or an assistant. Perhaps you can work with a friend who is also showing a dog, or with a neighbor or relative. (Many teenagers enjoy being given a chance to work with animals.) If your helper has an experienced eye, give him or her the clicker while you handle the dog. If your helper is not experienced enough to tell good movement from bad, have the helper run the dog back and forth while you watch and click.

The job of the observer is to click the instant the trotting dog happens to reach farther than usual with the front legs. (An easy way to spot even a small improvement is to crouch down to floor level and watch how far those front paws go in relation to the dog's nose.) *Click!*

The job of the handler is to trot the dog back and forth, and to stop *instantly* on hearing the click. The dog stops too, of course, and gets its treat. Then the handler resumes gaiting the dog. What if the dog doesn't seem to be extending at all? Then click and treat a few times at random. That makes the dog begin to feel "Hey! This is fun!" Then you'll likely see a new "spring" in the trot, giving you something to click. It doesn't matter that you interrupt the stride to stop and feed the dog; what the dog remembers is what it was doing when it heard the click. Be careful to click, stop and treat in the middle of the run, while the dog is traveling. Don't fall into the habit of waiting until the turnaround point, or you'll shape the behavior of lagging in the middle of the run and brightening up at the end.

With good teamwork, I've seen many dogs catch on to the extended trot, reaching farther deliberately, stride after stride, in two or three minutes; perhaps within a dozen clicks.

You don't need to worry about what the dog's back legs are doing. The dog will automatically engage its back legs more strongly to push itself forward more vigorously in front. You do need to watch for any tendency to "hackney," or raise the front legs high in a prancing gait; this is easy for dogs to do, and it's not what we're after. If the dog starts flinging its paws in the air, just don't click; the behavior will go away by itself. The handler should be careful, during this training, to keep the dog on a loose leash. A dog on a tight leash simply cannot move freely, much less learn a new movement. Even on the about-turns at the end of a run, encourage the dog to turn with you; don't spoil the fun by yanking it around. Meanwhile, the observer can help by being careful not to click if the leash is taut. With good teamwork, I've seen many dogs catch on to the extended trot, reaching farther deliberately, stride after stride, in two or three minutes; perhaps within a dozen clicks.

An extended trot calls upon leg and back muscles that a dog ordinarily may not use very often, so keep in mind that the dog may tire quickly. Tired dogs don't learn well. During the first week or two, plan to end each session—with a "jackpot" of a handful of treats—after a few good passes. Don't be tempted to ask for too much too soon, or the dog may come to dread this new task, a task that otherwise should be exhilarating and fun.

Watch your dog. For a day or two after the first lessons he may feel a bit stiff, just as you might if you took up a new sport. However, once learned, this behavior is a great muscle builder. A few bursts of extended trotting, every day or two, will do much to bring your dog into top athletic condition. That in itself will improve the dog's looks, the feel of its body under the judge's hands, and its general air of well-being.

As your dog learns to sustain the extended trot, and begins to understand what you mean by the cue, you can substitute a word—"Yes!" or "Good!"—for the click.

Adding the Cue and Dropping the Click

By withholding your click for progressively longer counts (five strides, ten strides, twenty) you can shape the behavior of sustaining the extended trot for a minute or more. You can also teach the dog to keep in stride around corners, as in the ring. Now you can add a cue: "Let's go" or "Show time" or perhaps a hand signal. Give your cue before you take off. Click (and treat) the dog after a few good strides, paying it off early, so to speak, for responding to the cue. Once the dog is springing forward when it hears the cue, you can go back to trotting for longer periods before clicking.

As your dog learns to sustain the extended trot, and begins to understand what you mean by the cue, you can substitute a word—"Yes!" or "Good!"—for the click. The click and treats are for teaching the behavior; they are the "language" you use to communicate what you want the dog to understand. Once the behavior has been learned, an occasional praise word or a pat will maintain it, and you'll need to get out the clicker again only if you want to improve the gaiting or repair some aspect of the behavior that has slipped in quality.

The Target Stick

A target stick is an easy way to cause extension, thus giving you something to click in your first lessons. Clicker-training suppliers sell folding aluminum target sticks, but you can use any stick or dowel about thirty inches long. Clicker train the dog to touch the end of the stick with its nose while walking along beside you. Move the stick here and there, clicking and treating, to shape the behavior of touching the nose to the tip of the stick no matter where that pesky stick goes.

(You can practice target training from your living room couch, by the way; you don't need to do it outdoors. And this exercise, described in more detail in Appendix C, comes in handy in many situations. For example, you can use a target stick to teach the dog to jump onto the grooming table, to get into a car or a crate, to retrieve selected items, and to do tricks such as closing doors. This is not a waste of time: your dog's brain needs exercising too!)

Once the dog has become infatuated with the target stick, take the stick along when you are gaiting the dog. As you trot the dog, move the tip of the stick out in front of the dog a yard or so. If the dog breaks into a canter, slow it down and try again. By and by the dog is likely to extend its trot to get to that wonderful stick. *Click!*

Another advantage of using the target stick is it allows you to place the dog exactly where you want it to be—out in front of you, say. When the extended trot has become reliable, and the dog is positioned well, you can reduce your use of the stick. You do this by replacing it with a verbal cue or hand signal, and clicking for the right movement and placement even though the stick is not there.

Limits

Like horses, dogs naturally vary in the amount of extension they can give you. The configuration of the shoulder is crucial; a sloping shoulder "frees" the dog's front movement, while a very vertical shoulder restricts the reach. Some breeds, huskies and dalmatians for example, are built to trot long, far, and fast, so many individuals that belong to these breeds can quickly learn to offer an extended trot. Other breeds, such as dobermans and some terriers, tend to have a rather vertical shoulder. This conformation may be correct for the breed, but it will result in a short trot. In most of the dwarfed or very short-legged breeds, such as corgis, basset hounds, and dachshunds, the extended trot is anatomically out of the question.

If you own a dog of one these short-legged breeds you might want to train for a high head and a happy tail, rather than extension, to improve your dog's movement in the ring. On the other hand, if you have a dog that is physically capable of flying like a ship in full sail, why rely on its dominant tendencies, or its feelings of the moment, to bring that behavior to the fore in the ring? Teach it the extended trot and you can guarantee a good performance, good feelings, and an admiring crowd as well whenever you hit the ring together.

CHAPTER 4

LITTLE DOGS WIN BIG

Shaping small dogs into show-stopping specials

Little dogs face challenges at conformation shows that big dogs do not. The length of the ring, which a big dog can cover in a few strides, can be a very long trot for a small dog. Bigger dogs are everywhere, posing a threat to even the bravest miniature pinscher. Worse, there are big people (with big feet) everywhere.

Small dogs know from experience that being stepped on hurts. I have often seen owners of little dogs waiting to go into the ring, oblivious to the fact that some bystander has just backed into or tripped over their West Highland white terrier or Manchester terrier, endangering and totally demoralizing the dog.

In addition, little dogs must be shown to the judge on a table, one high enough that the dog may be justifiably afraid of falling off. Getting there can be an unpleasant experience, too: with some terriers, it's traditional to hoist the dog roughly by the tail and the lead, an airborne experience that many small dogs visibly find disconcerting.

Training Dogs to Pose

Clicker training, the process of shaping behavior by means of conditioned reinforcer, offers enormous benefits to the small-breed conformation dog and its owner. Let's start with table work. One way to overcome a dog's fear of the examination table is to give it a job to do while it is being handled. Many handlers use a job—a behavior that captures the dog's attention—that can be described as "watch my finger."

To shape this behavior, get out your clicker (or a pocket stapler—anything that's easy to handle and makes a nice, distinct sound). Cut up some desirable food, such as hot dogs, chicken or cheese, into pea-sized pieces. Put the food where you can reach it but your dog can't—in your pocket or on a nearby stool. Put your dog on a table. *Click.* Give it a treat. Now hold your index finger out, steadily, in front of the dog's nose. Click when it looks at your finger, then take your finger away and give the dog a treat. You don't need to wave the finger and you don't need to have food in that hand; in fact, it's better if you don't—you want the dog to listen for the click, not look for the treat. (You're not tempting the dog with bait, hoping it will look interested; instead you're showing it that by focusing its eyes and ears on your index finger, it can make you go click.)

Sweet talk and encouragement may actually reinforce timid behavior.

If your dog crouches in fear whenever it is on the judging table, during your training sessions hold the lead for safety's sake, but don't push or pull at your dog, or try to lift it. Keep your hands away. After a few clicks for looking at your finger, your dog will stand up. Click the instant it stands, then give it a treat. Don't touch your dog, and whatever you do, don't talk to it! Sweet talk and encouragement may actually reinforce timid behavior. Instead, hold your finger enticingly near your dog's nose and rely on your clicker to tell the dog "That's what I want."

Your dog will break its pose while it eats the treat. That's fine. What happened when you clicked is what will count in the long run, not what the dog does between clicks while it eats.

Once your dog is standing and focusing on your finger, move it away slightly and start reinforcing your dog for leaning forward into a show stance. Keep the clicks and treats coming every few seconds in this early lesson by finding various good things for which to click: click for standing with all four paws on the ground, for leaning forward a bit, for raising or wagging its tail, and certainly for pricking up its ears.

What happens if you click for your dog's having pricked its ears and it simultaneously sits? Don't worry, give it a treat. The majority of clicks will catch your dog standing, and it's the cumulative information that counts, not the occasional mistake.

How do you extend the length of time your dog will stand on the table? Once you have the dog standing nicely, you can convey the idea that it should hold its pose by waiting two or three beats until you click. If your dog breaks the pose before you click, fine. Don't click. Start again. Let your dog discover for itself that it now needs to stand still a little longer to make you click.

Don't worry too much about the duration of the stand in the first session: a few seconds is a good start. Later, you can see whether your dog will hold the pose for twenty seconds or even a minute. At that point you can also begin to add distractions, such as working in strange places, in the presence of other dogs, or while a friend plays "judge" and examines the dog.

You are building a behavior you want, piece by piece, by using the click *to communicate what's right.*

In future lessons, you'll also want to click your dog for posing on the ground, thereby teach the dog to "self-stack" whenever you halt in the ring. In your first session, though, you should quit when your dog has learned to stand is beginning to pose. Take the dog down and let it digest what's just happened (while it digests its treats!).

What happened? This is the process known as shaping. Instead of physically manipulating your dog to put its feet and head in the right places, you are building a behavior you want, piece by piece, by using the *click* to communicate what's right. With a clicker and treats, you can get a pretty good stack from almost any dog in the first session, even if it's just a puppy. (Clicker training is harmless for puppies and they absolutely love it—see Chapter Sixteen for information on the benefits of using this method from the very start of a dog's life.)

Learning to Learn

You are now on your way to having a little dog that will stand with aplomb on a table while a judge examines it. But that's not all that's been accomplished. Your dog now has a job to do on the table, so it's no longer afraid, but busy. Your dog has learned how to do something easy and fun for itself. You have taught your dog to do what you want, but your dog thinks it's trained you to be predictable about producing treats. Your dog feels like it's in charge—no wonder it looks confident.

You also can use your clicking skills to improve your dog's gait, developing a more animated way of moving as described in Chapter Two. You can shape your dog to trot out a bit in front of you instead of at your side, which is very showy. You can give your dog the responsibility of holding its head up high, rather than you hauling it up with the lead. You can add charm, such a cocked head or pretty expression. One Lhasa apso with many Best of Breeds to its credit has learned to respond to the command "Say hi!" by waving its front paws. Judges always smile.

The purpose of clicker training a little dog is not to make a bad dog look good, but to give any dog the best chance it has in the ring. Mary Owens of Pensacola, Florida, has clicker trained Asti, her border terrier bitch, for obedience and conformation. Borders are rough-and-ready short-coated terriers, perhaps a bit unglamorous compared to the snappy wire fox terrier or the elegant Bedlington terrier. "But Asti knows what to do," says Owens. "She says to herself, 'Oh, show time again! Let's see, feet, tail, ears... and smile!'" What a pretty little dog. She's won Best of Group twice so far.

CHAPTER 5

PERFECTING THE SELF-STACK

Clicker train your dog to stack itself perfectly in the show ring

A great moment occurred in the Hound Group judging at Westminster in 1996. The judge daringly instructed the handlers to let their dogs stand freely, by themselves. Most of the dogs flunked, but when the handler of a beautiful Afghan bitch stepped back, leaving a loose lead between himself and the dog, she posed herself perfectly, as if to say, "Look at me. I can do this!" You bet she could. She won the group.

The self-stack, as this behavior is called, has lots of advantages. The dog voluntarily holds the position, so it appears happier. A self-stacking dog is always a crowd-pleaser; the dog looks good and seems smart. Judges are usually impressed. A perfect self-stack, however, is not accomplished in one or two clicker-training sessions. The stack is a compound behavior composed of may interlocking parts.

You have to build the self-stack unit by unit. Each unit or criteria is a separate shaping task. The process involves a certain amount of deliberate backsliding, a frightening thought to handlers who are used to correcting every mistake in an effort to "get the whole picture" from the beginning.

Positioning the Back Feet

Start training self-stacking by concentrating on the position of the back feet. Coax the dog into movement. Back up, encourage the dog to move a few steps toward you, then stop. If the dog by chance stops with its hind paws parallel to each other, click and treat. If not, try again. Some people prefer to walk toward the dog, causing the dog to back up instead of coaxing it forward.

Repeat until you have managed to click half a dozen halts with hind feet parallel, then do something else. (This start-and-stop business can be frustrating for the dog; five or six tries in the course of one session is plenty.)

If you keep clicking consistently for squared-up back paws, you'll soon get more and more square stops. In most cases, the back legs should be a hip-width apart, but some dogs look better with a slightly wider stance that lowers the rump a little. It's your choice; click for what you like. If you're new to showing, you might get a more experienced friend to watch the dog and click when the feet are properly set, while you handle and treat.

The dog's awareness of what it's getting clicked for will come in time. The dog may show that it understands by stopping with a foot out of place and voluntarily repositioning the foot. Click and give a big handful of treats if that happens.

Next you'll want to extend the time the dog will stand with its back feet glue correctly to the ground. Count three seconds to yourself, then click. Next time, count five seconds. Vary the time; working up to fifteen seconds or so. That's long enough to teach the dog that to earn a click it may have to stand still for just a few seconds or maybe for quite a while.

Front Feet and Weight Shift

When the back legs are fairly stable, start working on the front feet. With the dog standing still and its back feet placed properly, encourage the dog to move one front foot. Pull forward on the lead a little, tap the dog's toenails with your hand or foot, or lure the dog forward or sideways with a target or bait. You want the dog to move its front paws while keeping its back legs still. Exactly where the front paws go doesn't matter at first. Again, keep sessions short; quit when you've gotten in a few clicks.

When the dog is deliberately moving its front paws without moving its back paws, you can begin clicking selectively for movements that put the front paws in just the right place. The paws should be a shoulder-width apart and under the dog, not reaching out front like a rocking horse.

Meanwhile, keep your eyes open for a forward weight shift. Click and treat if the dog leans over its front paws. You can target the dog into leaning forward, but it may do so spontaneously. With the dog's weight forward, its hind legs will automatically be further back. The hind legs should be positioned so that the lower part of the leg is perpendicular to the ground.

Some people get the hind legs out from under the dog by physically pulling them further back. This can lead to an overstretched position that looks ludicrous and is uncomfortable for the dog. It's better to let the dog find its true stance by shifting its own weight, then the dog will be in the natural "alert and prepared for action" pose that is the goal of the stack.

Clicker trainers advise against using your food as bait to position the head. Use your finger or some other target to position the nose. Where the nose goes, the head will follow.

After the front paw position has been shaped, you can add a cue meaning "put your front paws where they should go." One exhibitor I know says, "Step, step." Another has taught her dogs to bounce their front ends into the air slightly on a lead cue: they naturally tend to land with their feet nicely placed.

Once you've shaped the hind paw position, the dog won't forget how it got those clicks. But now you're working on front paws. While the dog learns a new way to get clicked, it may temporarily forget to do the old behavior. That's fine. Just keep clicking until you're satisfied with the new behavior. Ignore other mistakes.

After the new unit has been built, start asking for good hind legs *and* good front legs before you click. The hind paw position is already in there somewhere; it'll come out again. Your clicker will "explain" that now, for a payoff, the dog must give you unit one plus unit two.

Head, Neck, and Ears

Next, study photographs of the desirable head position for your breed. Clicker trainers advise against using your food as bait to position the head. Use your finger or some other target to position the nose. Where the nose goes, the head will follow. Your click tells the dog when it's right.

Conventional handlers often put a thin collar right behind the dog's ears and pull upward on the lead to make the dog's head stay high. You can

shape a proud head carriage without that crutch by clicking for head height. It often helps to have a friend watch and click while you handle and treat.

When the dog is holding its head where you want it, click for what horse trainers call flexion at the poll. This is a little bend at the point where the head and the neck join. Big dogs and long-necked dogs look especially good with their heads held high.

To finish shaping the perfect self-stack, teach your dog to stand steadfastly. Push down on its shoulders and click for not moving. Push down on its nose and click for holding its head steady. The dog learns to stand still, but with its muscles "in gear." This will help it to remain motionless during the judge's exam and in the face of distractions.

Now you can add the final units: pricked ears, a kindly but determined eye, or a big smile, depending on the breed. Relax your standards on old units each time you start shaping a new one. Then put the units back together one by one; the process will go faster with each new unit.

By keeping the sessions short and mixing them up with fun stuff, you can train a dandy self-stacker in a few weeks and enjoy the satisfaction of having a dog that knows what it's doing, loves what it's doing, and looks its best in the ring.

CHAPTER 6

WHEN THE JUDGE ISN'T LOOKING

Games to play while waiting your turn in the ring

I once went to a dog show to watch a friend's smooth collie (it had a good chance of finishing that weekend). I ended up with a new appreciation of an aspect of show ring life that I have never considered before. Shetland sheepdogs were ahead of the smooth collies. A group of twenty-five Shelties entered the ring. The judge looked them over, one at a time. They were physically inspected, gaited away from and back to the judge, and gaited around the ring.

Big classes result in long waits. It was interesting to watch the dogs and their handlers lined up, waiting their turn: they were involved in a variety of activities, some more helpful than others.

Waiting in the Wings

Applied operant conditioning in the form of clicker training can help a dog look its best when it's being judged. But what do you, the handler, do when the judge is looking at other dogs? Do you try to keep your dog perfectly posed, just in case the judge glances your way? How long can you do that before your dog gets tired or bored? What if it's a large class?

Most of the handlers in the Sheltie class I was watching tried to keep their dogs stacked. They waved food around or tossed little toys until whatever interest their dogs might have had in the activity at hand was gone; when the dogs looked away, they would increased their efforts. Some handlers repositioned their dogs' feet repeatedly. One vigorously brushed her dog's immaculate neck fur while the dog flinched, its pained expression clearly communicating, "Enough!" A few handlers let their dogs go completely off-duty, and strain on their leashes toward the next dog. If the judge glanced at the line, was he seeing perfection, animation, and charm? Most likely he saw boredom, discomfort at holding positions too long (suggestive of back or hip problems), or downright misery.

During the long waits, use games that let the dog move around a bit.

So what *do* you do while waiting? Play games that attract positive attention without being obtrusive, that keep you and your dog relaxed and focused, and that show off your dog's strengths.

Playing the Waiting Games

Bouvier handler Susan Smith plays "Guess which hand?" with her charges. Stack the dog. Put your hands behind your back, then put both fists out, with a treat in one. The dog has to "point" to the correct hand without otherwise breaking its pose. If the dog guesses wrong, your hands go behind your back and then out again. The right choice gets a vocal "click" (you don't need to bring your mechanical clicker into the ring, just cluck your tongue against the roof of your mouth), then treat.

During the long waits, use games that let the dog move around a bit. "Where's the squirrel?" sends the dog to the end of its lead to find an imaginary squirrel. Ask a friend to stand a few feet away and tempt the dog with a treat or toy; when the dog goes "on alert," click and treat. A few sessions should be enough to establish the "on alert" behavior. Then give the cue "Where's the squirrel?" before your friend offers the temptation. Repeat in brief sessions, clicking for response, and you should get an end-of-lead eager pose with the verbal cue alone.

Here's an easy one: at home, teach your dog to catch a tossed treat. If your dog has a great front, bouncing into the air will bring its front feet down straight, drawing the judge's eyes to that feature. If your dog is a lousy catch, trainer Margie English suggests fastening its lead to something above its head, so it can't reach the ground with its mouth; then give it plenty of chances, but take back any treats that fall to the ground. Frustration will improve your dog's coordination.

Does your dog have good hindquarters? Doberman breeder Diana Hilliard suggests training your dog to bow. After the bow, when your dog stands up again for its treat, it should come back up with front and rear feet in line. If your dog has a superb rear articulation, emphasize the obedience kickback stand; from a sit, the dog stands from its rear end first, leaving its front feet in place.

Little exercises can also help keep your dog's mind occupied. Clicker train a wave, a soft bark, or a bounce. Catch these behaviors around the house, mark them with a clicker, and shape responses to your choice of cues. In the ring, ask for these behaviors in succession, alternating with other games. Remember, you don't need an actual clicker in the ring. Your "click" can be a tongue click or the word "Yes!" You might want to use hand signals rather than words, so you don't disturb your fellow exhibitors. You can reinforce with food, but once the behaviors are solid, eye contact and smiles will be all the reinforcement your dog will need.

You don't need an actual clicker in the ring. Your "click" can be a tongue click or the word "Yes!"

Discreet interactive play that is not wild and out of control can relax you and your dog while you're waiting for your turn. It attracts attention and increases crowd support. Audience response can help focus positive attention from some judges as well. And no matter what happens, whether you go home with a fistful of ribbons or not, you can say honestly that you *and* your dog had a great time at the show.

CHAPTER 7

LOOKING GOOD WITHOUT BAIT

Behavioral tips for helping dogs look their best in the ring

The judge was looking at a ring of eight male Ibizan hounds, all stacked in a line. The Ibizans I happen to know personally are laid-back, mild-mannered dogs possessed of a quaint sense of humor and two unusual physical characteristics. First, the long, whippy tail can hang loose in a soft S-curve, or, when the animal is feeling merry, it may be carried in a sort of snail coil over the animal's back. Second, Ibizans can lay their ears so flat on the head that they almost disappear, or prick them to attention like a German shepherd, in which case their ears are seen to be both sharply pointed and enormous, like a bat's. Just the thing for hunting small game in the desert, especially at night.

Naturally a handler with an Ibizan hound wants those gorgeous ears pricked and that long tail up and curled over the back. So every handler in this

It's much easier to clicker train the behavior of looking good, rather than trying to lure it into happening—and the trained behavior is more reliable.

class was working hard to make each dog look excited, or at least attentive. Some were enticing their dogs by waving food. Some were making noises with a squeaky toy. One handler was tossing a whistling toy with dangling streamers into the air. "Isn't this fun?" "Look, look!"

Meanwhile each Ibizan had his ears firmly folded away into his streamlined head, and his tail not just down but clasped between his hind legs so tightly that the tail's tip showed between the front legs. Every elegant face wore an expression of terminal disdain. It was not fun. They would not look. "Can we go home now, please?"

Even breeds that are easy to agitate, and that love the clamor of a dog show, can get jaded about being tempted with food and toys over and over again. The natural human reaction is to escalate the offer: better toys ("With streamers! That'll do it!") and richer food ("Liver!"). If—or when—that doesn't work, the next natural human reaction is to blame the dog. "He's getting stale," they say. Or, "He's ring-wise," or "He doesn't show well."

In fact, most often in such cases, the behavior of "looking good" has actually been *un*trained. The handler is promising something ahead of time (offering food or a toy a lot, giving it seldom) in the hopes of getting a behavior as a result: this is bribery, not training. The animal doesn't know what behavior is wanted, and even if it did, the bait and teasers may still be given or withheld unpredictably. No wonder the dog loses interest.

It's much easier to clicker train the behavior of looking good, rather than trying to lure it into happening—and the trained behavior is more reliable. My friend Lana decided to train her Ibizan hound, Bella, for the show ring, starting with the ears. We went to a public park. Bella saw a boy on a bicycle in the distance, and those big bat ears came up. Lana clicked and gave her a treat. Bella looked at the horizon again, and her ears came up even though the boy had gone. *Click.* Within a few minutes, Bella was pricking up her ears repeatedly. Now Lana held out her forefinger as a target for Bella to focus on. Bella looked at the finger while her ears were up. *Click.* "Looking good," Lana said, while her finger pointed at a spot a foot above Bella's nose. Bella looked and pricked up her ears. *Click.*

In one short training session, Bella had learned that pricked ears pay off. Furthermore she was beginning to learn a cue, or signal, for the behavior of pricking up the ears: the pointing finger and the words "Looking good." By repeating the experience in other locations and at other times of day, Lana was soon able to rely on Bella to prick up her ears whenever she said, "Looking good," and she was careful to click and treat whenever she did this, to keep Bella's trust in the cue.

In the actual ring, Lana used a praise word instead of the clicker, but she would always be careful to praise Bella when the dog's ears were up, and to give food as thanks for the right behavior rather than as a bribe beforehand. In competition, even if Bella was a bit daunted by all the strange sights and sounds, she would prick up her ears on cue (in fact dogs seem to gain confidence, during stressful moments, from being asked to do a behavior they know will be reinforced). As for the cue, the judge would not be surprised to see a competitor hold out a finger and tell the dog "Looking good!"—nor is a judge apt to notice that the dog looks better after hearing those words than it did before.

Targeting and Tails

Holding out a finger for the dog to focus on constitutes what clicker trainers call "targeting," providing a visual cue that the animal can look at, touch, or follow (see Appendix C). To get Bella's tail up over her back, Lana used a two-foot-long stick as a target. Bella already knew that she could get clicked for touching her nose to the tip of the stick. Now Lana gaited Bella up and down the path, holding the stick out in front of her so that Bella was running slightly ahead of Lana instead of beside her. The leash was loose; the target kept the dog in place. I stood to one side and watched. When Bella showed extra impulsion, or extra energy, I clicked. Lana stopped and fed the dog a treat. After three or four clicks, Bella began to enjoy the process; up came the tail. *Click.*

It didn't matter that Lana stopped, during that good gaiting, to feed the dog. The click came when the dog was moving well and the tail was up, so that's what was learned. Once the treat had been eaten, the dog was anxious to get back to the gaiting and show us a high tail and good motion again. There was also no need to haul Bella's head into the air by the leash, or to coax her to run fast; Bella was initiating good head carriage and good movement on her own, trying to earn clicks. In fact, soon Lana could gait Bella in a circle around her, head and tail up, the picture of animation, following the target stick—while Lana stood in one spot. *Click!* Add the cue, "Looking good" and the ears came up too. Jackpot! (A jackpot is an extra-big treat for a special achievement.)

The target stick is a special sort of cue, very useful in training, but not something one would take into the ring. A week or two later, when Bella's gaiting was reliably good, Lana would "fade" the target stick by slipping it backwards in her hand until her outstretched arm alone would be the cue for good gaiting.

Bella soon learned that if Lana held the leash loosely in her fist, at arm's length out in front of herself, Bella could get praise and treats for moving forward and showing herself well. She also learned that if Lana held the leash arm at her side, with her hand across her body, it was time to move close beside her trainer, in the obedience "heel" position. This latter arm position cue allowed Lana to maneuver delicate Bella safely at heel through the crowds around the ring and at the ring entrance without in any way taking the edge off her glamorous gaiting once they were in the ring.

There is never a penalty for failing to respond to a clicker-trained cue. If a dog doesn't prick its ears when you say "Looking good," the dog just doesn't quite understand that cue yet, or doesn't understand that it will pay off in a new environment or under new circumstances. Cues and targets are not "commands" in the obedience-training sense of that word. Cues are more like green lights for a particular behavior, and they are extremely useful once the dog understands and trusts them.

CHAPTER 8

CLICKING ACCORDING TO AKC RULES

How current obedience rules about training dogs on show grounds can help both dogs and exhibitors

In 1998, the American Kennel Club updated the rules in their Obedience Regulations concerning training dogs on show grounds. The old rules prohibited all training, and this excerpt suggests why: "There shall be no drilling nor intensive nor abusive training.... Physical or verbal disciplining of dogs shall not be permitted." The only exception was if a dog attacked a person or another dog.

Dogs quickly learn to identify the source of their own clicks—that is, their handlers—and to ignore clicks from other people and locations.

No practice rings were allowed on the grounds. Handlers were allowed to give only those commands necessary to move the dog about, and those commands could only be given in "normal" tones.

I guess these rules came about because some people were "training" their dogs harshly enough to upset spectators. But I can also see that the strictness of the old rules may have created unintended problems as well. After arriving at the show grounds, a dog that has been confined to a car or a crate may need to stretch and warm up with more than just a stroll. What better way to do this than with some quick recalls or a couple of heeling patterns? Oops! This would look as if you were training your dog, which was forbidden.

Now the rules have been loosened. You can warm up your dog with obedience competition exercises as long as you do them away from the show rings, keep your dog on a lead, and do not disrupt any other dog or person. This seems like a good compromise. Handlers can do a few warm-up exercises with their dogs without getting into trouble, but trainers who use severe corrections are still banned from using them on show grounds.

Making the Rules Work

The realization that reinforcement is still available in a new place and that the dog knows how to earn it is a tremendous confidence builder for dogs—and for people.

But what about the new wave of handlers who use positive training aids—toys, food and behavior-marking clickers—that some frown upon? These reinforcers are not permitted during competition, and some people believe that using them on the premises even outside the ring might be distracting to other people and their dogs, even if the user were some distance away.

In my experience, this is not a problem. Dogs quickly learn to identify the source of their own clicks—that is, their handlers—and to ignore clicks from other people and locations. Distant clicks outside the ring are ignored. Also, clicking and treats are for the learning stage of new behaviors. By the time dogs are ready to compete, competition behaviors should be polished, and the clicker should now have been replaced by a verbal marker signal—"Yes!" or "Good!"—to tell the dog it is doing a good job. Finally, by now any food treat should have been augmented or replaced by petting, play, or other positive reinforcement.

With a young or novice dog, I would run through enough behaviors on the show grounds so that the dog discovers that *our* rules (those learned with the clicker) are still in effect in this new, strange place. I want the dog to discover that reinforcement is still available and that correct behavior still pays off, even amidst new distractions.

Competitors should welcome the opportunity to warm up their dogs on show grounds, but one should never run through competition exercises in hopes of making last-minute corrections. Instead, use one or two of these exercises, and others such as "gimme five" or spinning, to verify for your dog that nothing has *really* changed.

Corally Burmaster, an Airedale terrier breeder and the editor of the *Clicker Journal*, once wrote about taking a young Airedale bitch to her

first show. An awning collapsed and frightened the puppy, who was then afraid to go into the show tent. Fortunately, the pup knew the behavior of targeting: she had learned to bump Burmaster's outstretched hand with her nose on the command "Touch." Burmaster held out a hand and said, "Touch!" The pup bumped and received a click and a treat. They did this again. She followed Burmaster's hand into the tent. The fear was forgotten and the success remembered.

Avoiding Hazards

I can foresee only one possible hazard to the current rules governing training on show grounds. One competitor proudly reported on the Internet that she had discovered that a clicker could be used effectively as a conditioned punisher. Once she had established the rule that a click would be followed by a "correction," she found she could stop her dog's misbehavior at a distance by clicking. Conditioned punishers are easy to establish and are taken seriously by their recipients, but the person who uses them may soon regret it. If the dog is in the ring and someone nearby even snaps a bottle cap, the dog will not wait to find out if the noise was intended for another dog; it might stop dead in its tracks or even bolt. Moral: Do not use conditioned punishers unless you are sure you have the only one around. (This may seem to contradict what I said earlier about dogs being able to ignore clicks from others, but once dogs are taught to respond to clicks as warnings of punishment, they are unlikely to take any chances about where the click originated.)

As both a trainer and a teacher, I applaud the current AKC rules. They give gentle latitude where it should be given. They also help make obedience competition less stressful and more rewarding for both human and canine participants.

CHAPTER 9

THE "KEEP-GOING" SIGNAL

If the click ends the behavior, how do you keep the behavior going?

In clicker training we often say "the click ends the behavior." This is certainly true in a simple shaping task. The dog sits. When it has been sitting for the right length of time, we click. Unless we give another cue immediately, the dog is then free to move.

Sometimes, though, when we're training more complex or long-duration behaviors, we need a reinforcing signal that does not end the behavior, but nevertheless tells the animal it's doing the right thing. Psychologist Joe Layng calls this an instructional or informational reinforcer, one that can be used to reinforce what is happening without also functioning as a termination signal. I call it a "keep going" signal.

How do you teach the meaning of the "keep going" signal? Simple, just start using it.

Emily Cain has trained and competed with many champion field dogs. Her current duck retriever, Jiffy, is a standard poodle (retrieving in water is what poodles were bred for originally). Jiffy was able to swim out and bring ducks (or targets) up to forty yards from shore. Emily needed to extend the distance Jiffy would be willing to swim, but splashing targets into the water at fifty or sixty yards didn't do the trick; Jiffy swam out about forty yards and then got worried and started swimming in circles.

In training a long-duration task like this one, we should deep the distances varied throughout the training. That is, as we gradually develop twenty-yard retrieves, and thirty-yard retrieves, now and then we throw in a two-yard retrieve, or a ten-yard retrieve.

Varying the distance—adding shorter as well as longer swims—has three benefits. First, the dog doesn't get discouraged by a task that always seems to get harder and harder. Even though, in overall terms, the task is getting harder, now and then it's really easy, and that reinforces the dog's willingness to try. Secondly, the dog becomes "fluent" at the task: it learns that ducks can fall close, or they can fall far, and that if you don't find it where you thought or hoped it would be, keep looking. Third, in any given training session the dog achieves more successes and reinforcements with less fatigue, making your training time more productive.

However, human nature seems to make us want to train from "easy" to "hard" in a straight progression. Maybe Jiffy's behavior is starting to extinguish because of the steadily increasing difficulty of the task. Or maybe Emily varied the distances sometimes but also reinforced Jiffy enough times at or about the forty yard zone that Jiffy presently considers that to be the limit. Or, or, or…

How this impasse came about is not important. What can we do about it now? We might click when Jiffy is swimming straight and before she starts to circle when she gets past the forty yard mark, and then let her quit and come back for her treat. In this way we could increase the distance over which she will swim straight.

However, if she has to come all the way back to get reinforced, she'll get tired twice as fast. Maybe she will get tired before she catches on to the idea that swimming further is a good thing. Well, we could put a boat out there and reinforce her from the boat and then send her onwards again. But she might start considering the boat as a cue or a target, and fail to swim farther than usual when no boat is present. Here's where a "keep going, you're on the right track" signal would be handy.

The signal needs to be different from your regular click, though. Why not human words, encouraging cheers, say? Too vague: dogs hear so many words, and words shouted at a distance may convey what you wish, or they may not. An arbitrary sound is better. A whistle can serve, perhaps a special whistle that's used just for this purpose. Or try a special signal on a regular sports whistle: short-long-short, say. Just pick something different from any regular signal you might use, such as the two short blasts that are the standard field signal for "Come back home."

How do you teach the meaning of the "keep going" signal? Simple, just start using it. Use it first during easy continuous tasks that your dog already knows: a one-minute long down, say. Shortly before the end of the time, you give the "keep-going" signal. Then you return to the dog, click, and treat. Or, while the dog is performing some other job (a dry-land retrieve, for example), toot when the dog picks up the object—"That's right, keep going"—then click and treat as usual when the dog gives the object to you.

Because this signal becomes information—real, useful, beneficial information—it also becomes reinforcing in and of itself.

You never need to pair the "keep going" signal with a primary reinforcer such as food or praise. The dog will discover, on its own, that the keep-going signal means a click lies ahead, somewhere. Just by using the sound, you turn it into information.

Because this signal becomes information—real, useful, beneficial information—it also becomes reinforcing in and of itself. Therefore, with care, you can use this informational reinforcer during an on-going behavior to reinforce or strengthen behavioral details—such as foot placement, holding without mouthing, or anything else—without interrupting the main behavior at all.

Steve White, a police officer and former head trainer for the Seattle Police K9 corps, uses an informational reinforcer to help dogs learn to track. When a dog is tracking, you don't want it to stop for a primary reinforcer until it reaches the goal at the end of the track, but it is nice to be able to help the dog a little.

Here's one way Steve does it. In the early stages of training, the handler knows where the scent track has been laid down, but the dog doesn't. Suppose the dog overshoots the track, and becomes confused. The handler watches and waits. At the moment the dog shows signs of having picked up the scent again, or is at least heading in the right direction, the handler gives the 'keep going' signal. You can see the dog relax a little—"Oh, okay, I guessed right"—and keep on tracking correctly. Steve's trainers use a muffled click as their informational reinforcer: the handler wraps his hand around the clicker and depresses it with the middle finger, making a sort of cluck-cluck sound that the dogs easily distinguish from the sharp "Job's done!" terminating click.

Can you use the informational reinforcer more than once, during the same task? Of course. One must be careful, however, not to use it so many times before the click arrives that is ceases to be a promise and becomes a lie. Also, as with overuse of any reinforcer, frequent repetitions of the 'keep going' signal can lead a dog to think it must be doing the wrong thing whenever those signals stop. Then at the first slackening of the rate of the signals, the dog may change its behavior, or even quit altogether, figuring the game is over.

You often see this byproduct of a too-frequent or nearly constant stream of reinforcers in dogs that have been trained with food treats, but without a clear-cut conditioned reinforcer. As soon as the food stops coming, the dog stops working—not because it is "food crazy" but because, without a conditioned reinforcer, the message is in the food itself, and with the food taken away, the message is gone and so the game must be over.

Here is another twist on the informational reinforcer. Reinforcement expert Bob Bailey, while training dolphins for the U.S. Navy, used a continuous tone, delivered by a collar or harness, to steer dolphins many miles in the ocean. If the dolphin deflected from the correct line, the tone stopped; the animal then swung left and right until it found the

tone again, then went on with confidence. Bailey's Navy dolphins could be relied on to work for hours and many miles before receiving their primary reinforcers when the task was over. If you think that would only work with a dolphin, take note: Bailey later used the same system with house cats, and was able to steer cats through city streets and public buildings, for hours, for an eventual food reinforcer.

And where does the "keep going" signal fit in the conformation show ring? To begin with, it's extremely useful any time that you need your dog to keep doing what it's doing for an unusually long time. For example there usually comes a moment when the judge wants to look at all the dogs, all stacked and posed, together. If there are five dogs, the judge's perusal of the line takes a minute or less; but if it's a really big class, the dogs might have to hold still for an uncomfortably long time. Here's where a well learned and well understood message conveying the idea "Hang in there, you're doing great," can keep your dog looking focused and happy instead of fidgety and miserable, no matter how long the judge needs to study the class as a whole. If you have created and established the "keep going" signal under other circumstances, for example when practicing gaiting, or when exercising the dog to build muscle and endurance, you'll have a valuable tool whenever any long-duration behavior is called for and the dog needs help.

CHAPTER 10

CHANGING THE WORLD, ONE CLICK AT A TIME, PART I

A master-class clicker training adventure

In the fall of 1998, a dog trainer in my Boston neighborhood, Mary Ann Callahan, approached me about giving a clicker training class for serious trainers like herself. The class would be based on my book *Don't Shoot the Dog*, a text on the scientific principles underlying operant conditioning with positive reinforcement, or clicker training. Despite its title, the book is not just about dogs, but about all kinds of applications of positive reinforcement, but many dog trainers find it fundamental to their work.

I had been teaching occasional two-day seminars to large groups of dog trainers, but I had never really taught a continuing clicker class with the same students week after week. I accepted Mary Ann's proposal and asked for a group of eight participants who were already familiar with clicker training using the method in their own competing or teaching. Mary Ann found the participants by word of mouth and organized the schedule. Originally I planned to run the class for six weeks. Somehow it got so interesting that we stayed together for nearly a year.

Clicking on Time

Instead of tackling training tasks one behavior at a time, as is customary in dog training classes, I decided to address specific clicker training skills, or fluencies. For example, did everyone understand how to click at the right instant? The click is a marker signal. The click promises a reward to come, of food or petting and praise, and it also identifies the behavior that earned the benefit; this is its most important function. To work as a marker signal the click needs to 'fire' exactly when the dog is doing what you like: while the behavior is happening, not after the behavior is over (which is when one gives praise or treats).

An exercise in 'hind-end awareness,' based on the teachings of Linda Tellington-Jones, proved to be a great way to sharpen the trainer's timing skills. We spread out a lot of objects to use as low obstacles: a ladder lying flat, a vacuum cleaner, a coil of rope, a few boxes. Then we led each dog across each object. The trainer's task was to click exactly when the hind feet crossed the obstacle, but ONLY if they crossed without touching it. If the dog was a little clumsy, the trainer might have to click just for one foot at a time, training the left and the right hind legs in separate tries.

In a few tries we could all see the vast difference between clicking and then rewarding a good performance after it happened, even if only half a second *after* completion, and clicking *during* the crucial behavior, and then giving the treat. Dogs that were clicked and rewarded after completion were eager to try again, but kept on making mistakes. Even though they got the same praise and treats on the other side, the dogs that were clicked *while* their back legs were moving over the obstacle did much better. They quickly learned, often in two or three clicked trials, to step carefully through and over things, instead of just stumbling or crashing along any old way. This exercise made the dogs much more attentive to their own actions, a nice skill for the dogs. It also made the clicker-wielders much more attentive to the timing of the click, a nice skill for trainers.

Team training

Another necessary skill for clicker trainers is planning a shaping session: figuring out how to break a behavior down into many small steps, and then training it step by step. Traditional trainers often try to 'go for the whole picture' from the beginning. But if the dog doesn't do the whole behavior right, you have nothing to reinforce. People tend to get stuck here, or to fall back on correction to try to build the behavior. I used team training to break up this logjam.

I started by having each person show us a behavior they were having trouble with at present. (Maybe the dog was giving good 'fronts' but too far from the trainer; or dropping the dumbbell; or missing a weave pole; or rolling over only halfway and then rolling back.). Then we divided up in pairs: two trainers, two dogs. Each team crated, tethered, or downed one dog. Then the person without a dog clicked the other person's dog through whatever behavior the owner wanted to work on, while the owner handled the dog and gave out the treats. The teams had ten minutes to train the first dog. Then they changed places and worked with the other one.

A partner fresh on the scene can often see new ways to break down the behavior that's perplexing you into smaller steps, and that's what happened with our teams. Having the owner do the handling and feeding made the situation familiar for the dog, so the dogs cooperated calmly. Soon each dog showed a better understanding of what was expected: Fronts have to be close to the owner to earn a click; rolling all the way over is the winning thing to do. The improvement was sometimes stupendous, and the human members of the class often reinforced each other with cheers, laughter, and applause. It was fun for the participants, the spectators, and of course the dogs.

Latencies

Often a dog understands what is wanted, but takes a long time to do it. This is what behavioral scientists call 'a long latency.' Latency is the time that passes between giving the cue ('Down') and the dog actually getting to the floor. Good trainers aim for 'zero latency,' i.e. the dog is in motion before the hand signal is completed or the words are out of your mouth.

You can build short latencies with the clicker, by selectively clicking for faster responses. Begin by asking for several repetitions of the behavior. Click and treat every response, so the dog becomes enthusiastic about the task. Then ask for several responses again, but only click about half of the responses—the quicker ones. Jackpot—give an unusually big reward—for any truly quick response. By this method, that leisurely, oh-my-aching-back 'down' can turn into an instant belly-flop down in less than five minutes. We worked on fast responses on lots of behaviors: fast sits, fast recalls, and instant downs, including down at a distance or in mid-stride. If you make quick response a non-negotiable criterion for everything you click, dogs and trainers both start getting pretty speedy.

Real-world benefits

As the classes progressed, we all found that each new thing we and our dogs learned proved to be useful in unforeseen ways. No matter what kind of work the dogs were being asked to do at home, we were seeing improvement, sometimes with no specific training effort at all. The class participants participated in a wide variety of canine sports, including obedience and agility competitions, herding, tracking, carting, and the conformation show ring. For dogs being trained for the breed show ring, for example, hind-end awareness, developed by clicking for crossing little obstacles without touching them, turned out to be very beneficial in training the tail set. Dogs often don't know they have tails, and at first can't control them consciously. If you click a dog for holding the tail higher, its first conscious efforts to move the tail result in wags, not lifts! Going through the Tellington-Jones exercise made the dogs of where their tails were, which made it much easier to teach a dog to carry the tail correctly (either higher or lower, as required by its particular breed). Being aware of back foot placement also facilitated training the self-stack; and the habit of training for 'zero latency' resulted in quick response to any and all cues. As we continued, over the next months, we were to find many more practical applications of our experimental work, described in the next two chapters.

CHAPTER 11

CHANGING THE WORLD, ONE CLICK AT A TIME, PART 2

Inventive clicker training can lead to success in many training areas

Each week with my master clicker training class was a new adventure in shaping behavior, in building it step-by-step, in small increments. Here are more of the exercises we did as we all learned more about operant conditioning—the art of training behavior without using force or punishment.

Targeting

One of the exercises we worked on repeatedly was teaching all the dogs to target (touch and follow) the tip of a stick with their noses. Before our first targeting session, I stopped at a hardware store and bought a dowel for each member of the class. The dowels were three feet long and about a half-inch in diameter. We spent our hour together teaching each dog to touch the stick with its nose, then to follow it to the left, to the right, up, down and forward. Some owners and some dogs caught on immediately, while others had to try several approaches before they got it right.

Target training (outlined in detail in Appendix C) is a great behavior to try if you want to learn clicker training. First, there is no way the trainer can make the dog touch or follow the stick: the behavior only happens if the dog does it voluntarily. Second, this exercise is one in which you can easily see how the behavior is "shaped," starting with tiny movements of an inch or so, and gradually building up until the dog is following the stick for long distances.

"Pass" the Dog

It is amazing how dogs know which click is meant for them, no matter how many people are in the same room, clicking more or less simultaneously.

For this exercise, we divided the group into pairs of trainers. Using our target sticks, we tried "passing" each dog from one person's target stick to the other person's and back again. Once the dogs caught on, each team could do figure eights. The dog would circle one person, then pass to the next target stick and circle the other person in the other direction.

The next exercise we worked on was passing the dogs down a line of three or four people, having them weave in and out as they went. It turned out that every person had a distinctly personal shaping style. What you clicked was what you got. (It is amazing how dogs know which click is meant for them, no matter how many people are in the same room, clicking more or less simultaneously.) If you were in a line, you had to shape your section of the passage, building reliability and duration, just as if you were the only one working the dog.

For us humans, this was a *big* lesson in thinking on our feet and monitoring our own shaping techniques. Suppose a particular dog always wanders away from your target, but it sticks like glue to the targets of the people on each side of you? You know right away that it isn't the dog's fault, and it isn't chance or bad luck; you have done something incorrectly, probably involving the timing of your clicking.

It happened to me. One student, Dee, was working a brilliant little Border Collie, Nepe, a herding champion despite her tiny size. Nepe followed my stick twice, even though I did not click, and then left me flat on the third pass. "Hey, twice with no clicks? Forget it."

"Oops," I said to myself. "Because this dog is so smart, I figured I didn't have to reinforce it! I was wrong!" This sort of thing happened to most of us at some point, and we learned to think fast and fix the problem in a couple of clicks so the game could go on.

We probably spent no more than three twenty-minute sessions on this exercise, over the course of two or three classes. But many dogs ended up being able to weave down and back a six-person line in order to receive one treat at the end. And none of us worked on it at home.

Left- and Right-Directed Go-Outs

When quick responses, fast clicks and an understanding of how to shape behavior in increments were well-established, we could try some tougher problems. We decided to try to shape the behavior of going away from the trainer, not just straight ahead but also on angles in an indicated direction. Each person had twenty minutes in which to shape his or her dog to go away from the trainer at a 45-degree angle to the left or the right. I gave no suggestions or instructions about how to do this.

Almost every dog-human pair got at least one direction down pat in their twenty allotted minutes. What was interesting, however, was that every trainer came up with a different way of doing it. The final show-and-tell was less about what the members of the class had accomplished than it was about how they had arrived at that behavior. By tossing a lure? Working against the wall or in a corner? Cold-shaping, paw by paw, step by step? Using a chair as a go-around obstacle? Targeting to two objects on the floor? Using a human partner? Or two? And when did you click? This last was as important a question as how one had shaped the result.

Tricks

Homework was not usually a part of the class. I wanted people to learn more shaping skills, but I wasn't interested in monitoring or reinforcing specific achievements or performance. I did not want to impose duties of any sort or keep track of assignments. I was also anxious to avoid creating any kinds of competitive feelings in the class.

We did, however, have "trick challenges." Anyone could bring in a new trick, and everyone who wanted to try had a week in which to train it. Israel trained his big, sweet-natured Doberman, Knight, to fake a snarl

on the cue "Say Grr." Dee developed a great cue; whenever she sneezed, her Border collie Nepe pulled a Kleenex out of the box and brought it to her. Inspired by this, several of us tried to get a dog to sneeze on cue (so another dog could bring it a Kleenex); that turned out to be much more difficult!

Benefits Outside of Class

Along the way, we began to see results outside of class. Kevy, the Irish Terrier, qualified in agility for the first time. Nicky, a Keeshond, got his CD. Tucker, a beautiful and very laid-back Bernese Mountain Dog, passed his carting tests. Benny, a belligerent Golden Retriever, who at first openly displayed an intense desire to attack any other dog that walked by or even looked in his direction, became serene enough to attend and compete in obedience trials, and earn his first titles. He was a transformed dog, according to those who "knew him when."

Did we specifically work on Benny's aggression? Not as a rule, although in one class, toward the end, we asked all the dogs to jump over other dogs, and also to lie down while other dogs jumped over them. That exercise was especially good for the naturally pugnacious terriers in the class, and for Benny, whose aggression came from fear. It was no trouble at all for the benign Bernese mountain dog, Tucker, who couldn't care less how many dogs jumped over him.

In fact I was seeing new confidence in all the trainers and all the dogs. Perhaps the clicker training process brings humans and animals alike a new view of life: as an ongoing opportunity for making good things happen, instead of as a threatening world in which one must constantly strive to prevent bad things from happening. In the months and years that followed, several class members took their day jobs into totally new directions, or expanded their dog training skills into careers. Inventive clicker training seemed to be leading to success in many areas, not just in the ring.

CHAPTER 12

CHANGING THE WORLD, ONE CLICK AT A TIME, PART 3

Cues as clicks, and the final results

As our master class wrapped up many months of exploratory clicker training, we tried a final exercise that any dog owner might find interesting to do with their own dog.

In the first part of the exercise, I asked everyone to pick a short, simple behavior that his or her dog loved to do and that was under good cue control. Sandy, for example, had gotten rid of her border collie's excessive barking by bringing it under stimulus control. Now her dog, Jolie, was silent, unless she heard that wonderful cue, "Bark," which gave her the opportunity to get a reward for what she most loved to do; so for the purposes of this exercise Sandy's choice was the behavior of barking.

Some of the other choices made by members of the class included a spin, a high-five, a recall and a bow; anything quick and easy that the dog did very reliably on cue. Each person then demonstrated for the class that the dog understood the chosen behavior well and did it with gusto.

We then set up various obstacles that people had brought in for the class. Some were borrowed from agility training and some we came up with on our own. I asked the members of the class to choose an obstacle their dog had never encountered before. I specifically told them to choose something the dog might find a little bit daunting, for example, pushing through a curtain that the dog could not see past, standing on a wobble board, or jumping through a smallish tire.

Each member of the class used a clicker to shape their dog's navigation of the chosen obstacle, while I walked around and watched. I was waiting for the point at which each dog had the idea and was beginning to do the new task although hesitantly. All eight dogs reached this stage in about 10 minutes.

Then I stopped everybody and told them, "From now on, as soon as you get an approximation of the behavior—two or three paws on the wobble board, say—instead of clicking, please give that favorite cue which your dog loves. Then let that familiar behavior happen, then click and treat that behavior." One by one, the trainers each used their dog's favorite cue as the marker signal for trying the difficult task. In other words, the cue word was to function as a click, a signal that identifies to the dog that the behavior it is performing at that moment is

the behavior the trainer desires. The cue also promises that a reward is coming. As each dog tackled its obstacle, it heard the cue, responded with the bark, spin or other familiar behavior that it loved to do, and then got clicked and treated. Suddenly I was hearing cries of amazement all over the room. After two or three repetitions of this sequence, each dog was doing the scary, previously resisted behavior perfectly, and with gusto.

What was happening? At least three factors were at work. We had built a behavior chain, ending up with a well-learned behavior, so the chain itself was reinforcing. Also, because the known, cued behavior was a favorite, we were using the Premack Principle. (David Premack, a psychologist who works with primates, was the first person to write about the effectiveness of using a preferred behavior as a reward for the performance of a disliked behavior.)

Most important, we were using a favorite cue both to identify *and* to reinforce the new, more difficult behavior. It was a cue *and* a click, melded into one. Each dog got two rewards: the chance to do the behavior it loved, and a treat. What especially surprised everyone was how quickly the dogs learned to do the scary thing to get the good cue. The old cue was actually more powerful than a simple click and treat.

The timing of the cue/click had to be perfect. Sandy had to give the bark cue to her Border collie at the instant her dog lifted its fourth paw onto the wobble board, exactly as she would have timed a click. But then—what power! Wobble boards lost their terror for the dog on the spot. Instead the task represented a great new way to make Sandy give the bark cue. It was a fabulous new tool.

The class members had now experienced the process personally, not only by doing the training, but by experiencing the thrill of their dogs' successes and their own astonishment at the disappearance of fear and reluctance. Perhaps because of the emotional as well as the intellectual experience they had just had, it was a tool they now could really understand and apply. The trainers plunged into an excited discussion of what they could do in other training situations using this new "super click."

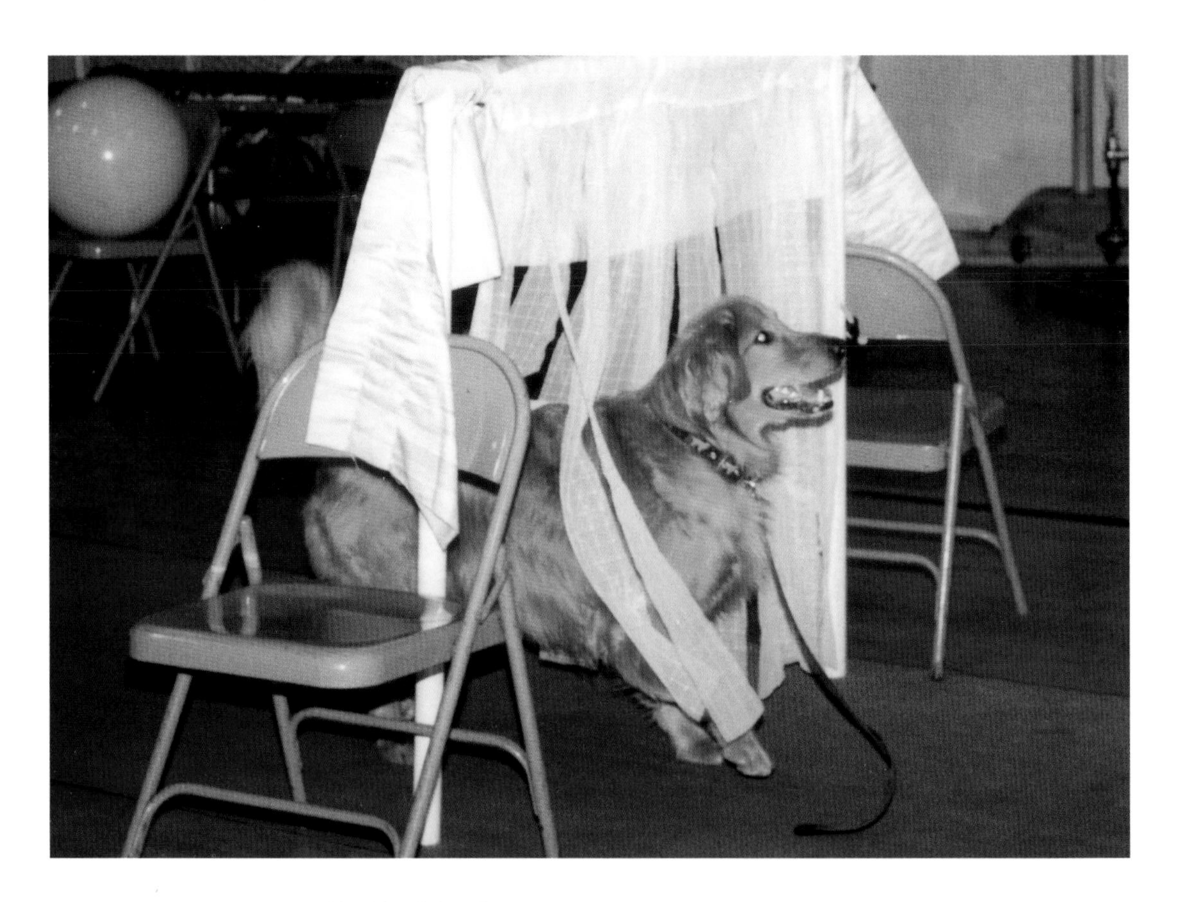

The Final Results

Every dog was doing things it never had done before.

On the last evening of the master class each trainer made use of the chance to repeat the intriguing "cues as clicks" exercises with a new obstacle. Every dog was doing things it never had done before. The golden retriever moved backward through a curtain, the poodle stood happily on the wobble board, and the Irish terrier learned to score soccer goals with its front legs. All the owners were beaming at their dogs, and every dog was beaming back. Our class had become a weekly hour of sheer, fascinating joy for every living being present.

But what did we truly accomplish? I remember one evening when we were in the middle of trying out some new challenge, I stopped and made a confession: "You know, I don't really know what we're doing here."

Class member Israel Meir instantly answered, "Changing the world, one click at a time." So we were. And so can you! Click!

CHAPTER 13

BEGGAR'S HOLIDAY

Training your dog to take a permanent vacation from begging will help it grow up

We have all experienced dogs begging—if not our own, then our friends' dogs. We know the moist-eyed stare, the importuning paw, the heavy head on our knee under the table, the soft whining. We know the small dog's leaping, lapping, yapping, and nipping that says "*Gimme, gimme, gimme.*" These behaviors compel us to respond with attention, food or whatever it takes to stop the begging.

Begging begins as a natural puppy behavior. Canid puppies of all species greet canid adults and coax food from them by licking their faces, jumping, whining and begging. But when an adult dog in a human household begs—particularly when the begging is frequent, persistent and annoying—it is exhibiting owner-trained behavior.

Why Begging Persists

Why do people give in to begging? Some do so because they actually like the behavior. The puppylike behavior and soulful looks trigger a response that is similar to our innate response to infants: "Aw," we say, "isn't that cute!"

Sometimes people give in because they read a dog's begging behavior as affection. I think that is why some people secretly feed a dog under the table, even though doing so is against household rules. They feel as though they have a special bond with the dog because it risks punishment to seek them out. The dog's behavior, of course, is reinforced by the treats, and the surreptitious giving of treats is reinforced by the dog's soulful eye contact and repeated visits.

Begging becomes persistent—and escalates to the level of being infuriating—not because it works, but because it works only some of the time.

Begging becomes persistent—and escalates to the level of being infuriating—not because it works, but because it works only some of the time. If owners responded either immediately or not at all, then every puppy would learn early on that if you beg and the answer is no, you might as well quit. This is how adult dogs, wolves, foxes, and other canids teach puppies to stop begging. They react immediately either with a *yes* (by regurgitating or dropping food in front of the pup) or with a *no* (by turning away, leaving or, if necessary, reprimanding the pup).

Many people, on the other hand, are more virtuous. They tolerate the begging, they try to ignore it, they explain to the dog why they are not going to feed it right now. Then, when none of that works, they give in and feed the dog.

Thus the dog develops the skill of begging vigorously and for prolonged periods. The owner theoretically knows better than to give in, but eventually does just to buy peace. Like feeding the beggar under the table, this is a symmetrical interaction. The eventual cessation of the disturbance reinforces two of the owner's behaviors: giving in and tolerating longer and longer periods of begging. (The same mechanism, technically called a random schedule of intermittent reinforcement, maintains whining and tantrums between human parents and children, often for years and in some cases for generations.)

Peculiar and individual begging behaviors can also be reinforced without owners having any idea that they have created the behaviors in the first place. I once visited a kennel of West Highland White Terriers in which barking for attention had reached such intolerable levels that all the dogs, even the six-month-old show prospect, had been surgically debarked. That didn't stop the behavior of barking, which still soared whenever the owner looked at or moved toward the dogs; it just reduced it to a peculiar rasping sound. Some of the dogs, including the puppy, had also developed a rapid spin to the right, which they executed whenever it looked as if their door might be opened. The breeder though it was genetic. "Her mother did that too," she told me in wonder, as the puppy whirled skillfully around and around, hoping to be released. Maybe her mother did do it, but it was the owner who unwittingly created these begging behaviors in all of her dogs.

Except for the fact that begging is a nuisance, what is the harm? Begging is commonplace for all domestic animals. Horses nuzzle their owners for treats; cows moo at the gate; sheep baa for fodder. Social worker and clicker trainer Lynn Loar says, "Animals are bred for their juvenile qualities and physical characteristics, and they get reinforced for them as well. Begging is a normal infantile behavior that sticks around longer in domesticated animals because of husbandry practices and the ability of begging to propitiate the owner."

Maturity, even for dogs, requires the ability to act independently, to solve problems and provide answers instead of questions.

Bringing Up Baby

However normal it may be, begging is dependent, immature behavior. Maturity, even for dogs, requires the ability to act independently, to solve problems and provide answers instead of questions. Some people, of course, like their dogs to remain immature: "She's my *baby*," people say. But owners and dogs for whom begging is a primary mode of interaction are missing out on a much richer level of communication—and a much stronger bond.

Whatever the species, the animal that begs has not yet learned that its own mature, independently offered, learned behavior can cause good results, too; so it uses infantile behavior—nuzzling, crying, pawing, nipping, etc.—to try to get what it wants. How do you help your dog grow up? Any non-punitive training that develops initiative gives the animal an alternative repertoire; flyball or agility competition would be ideal. Or try some indoor clicker games, such as hide and seek or finding a hidden toy. Clicker training is especially useful for the pet owner, because it helps you notice behavior you like, and reward that behavior with your treats, instead of unintentionally rewarding behavior you actually wish would go away. And an animal that has learned some real things to do can indicate that it wants treats by offering its own working skills, instead of by babyish begging. That's lots more rewarding for the owner than the small pleasure of being obliged to share one's breakfast just because the dog looks cute.

CHAPTER 14

ALTERNATIVES TO THE CLICKER:

A clicker is not the only device that works effectively as a marker

Wow! After years of training animals using a conditioned reinforcer, I look around and, suddenly "clicker training" is all the rage. People are using clickers to tell agility dogs when they've correctly hit the contact zones or entry and exit points for each obstacle on the course. They're using clickers to fine-tune straight sits and fronts and heeling position in obedience dogs. They're taking clickers into the conformation ring. The clicker, with its pinpoint accuracy of communication and its message of positive reinforcers to come, speeds up the training and makes it fun for handler and dog alike.

But sometimes people confuse the message with the messenger. They feel they can't train this way unless they locate a "real" clicker, the oblong boxes enclosing a bit of spring steel. But do you have to use a clicker? Is there something magical about that device?

Certainly not. You can make a suitable sound with a ballpoint pen or a pocket stapler or a Snapple bottle cap. Writer Josh Adams reports that his dog responds enthusiastically to the sound of shutting a hair barrette. Or you can use what dolphin trainers use, a whistle. I use a clicker for both my dogs, but a little tin whistle when I want to train just the pup. That way the older dog doesn't drive us both crazy begging me to train her instead of him; she doesn't associate the whistle with anything at all.

The one thing that does not make a very good substitute for the clicker is a spoken word.

And who says the marker signal has to be a sound? What if you're training a deaf dog? Behaviorist Gary Wilkes uses a fast thumbs-up gesture or an "Okay" signal. Lin Gardener gives school demonstrations with her stone-deaf rescued Australian Shepherd, Maggie Mae. Maggie Mae recognizes over eighty sign language cues; and her "click" for doing the right trick is the blink of a flashlight.

Any sense will serve; you can even use touch. At a professional meeting of behavioral psychologists I watched trainer Mark Lipsitt give a brilliant demonstration of 'clicker training' with a young field-bred Labrador Retriever, using a quick tap of his hand on the dog's cheek as the click. The psychologists watched the dog, heel, sit, stay, recall, and so on, with no leash, and no praise or correction, and many commented that this was a very smart dog. Wrong: the dog was average; the trainer was smart. None of the observers had connected the swift touch to the dog's face—and the dog's joy when it got the tap—with the dog's trained behavior.

Why words don't work

The one thing that does not make a very good substitute for the clicker is a spoken word. Why would that be? Dogs love praise, don't they? The problem is, praise is a good reinforcer, but it is not a good marker signal. First, we talk all day long; a marker signal needs to stand out, but any spoken word just tends to get lost in the clutter. Second, most words take too long. Say "Good dog!" for coming when called, and the dog can be going away again before you get it out.

Some people choose to use a short word as a marker: "Yes!" for example. That will work fairly well as long as you are careful not to use the word casually (never saying "Yes!" again when you are not addressing the dog) and as long as you remain consistent. But how often do we say "Yes" sometimes, and "Attaboy" or "That's right!" at other times? Finally, whenever we speak we unavoidably broadcast our emotions. As far as my dogs are concerned, even my sweetest praise words are bad news if I'm irritated. But no matter what mood I'm in, my click is the same good news every time.

The click benefits the training in another way. Because the click is arbitrary, it provides you with feedback that your own voice just can't do. One of the biggest problems in pet dog training classes is poor timing. Beginners don't easily recognize that their spoken praise or corrections have come far too late to do any good. But the click is different. If you want the dog to sit, but you click as it stands up, you *know* it! "Oops, I was late!" The click teaches timing to the trainer, as well as teaching behavior to the dog.

An acoustic arrow

There is also, I suspect, a physiological reason why it's actually best if the marker signal is something mechanical. Research on fear conditioning, in both animals and humans, indicates that certain kinds of sharp, sudden stimuli are processed first by the amygdala, in the oldest part of the brain, before going to the conscious mind. This old part of the brain also gives rise to emotions. If a certain startling sound has been associated with a specific event in the past, the sound is instantly classified either as scary—"Yikes, let's get out of here!"—or exciting: "Oh boy! Good news!"

A spoken word is almost never going to have that primeval power. Words must be identified and interpreted by the brain before causing a response. A brief, sharp mechanical sound, however, is always going to hit the oldest part of the brain first, and set off the emotion—a positive thrill, in the case of the marker signal—that makes clicker training so exciting for the trainees.

I compare the clicker to an acoustic arrow that goes straight into the animal's nervous system. The dog doesn't *have* to think, in order to understand the click. The marker signal is an incredibly powerful tool, and pet-friendly, too, since it always means good things. But what kind of signal is best? Clicker? Whistle? Flashlight? Buzz? That's up to you.

CHAPTER 15

WHAT THE CLICK ISN'T:

Common misuses of the clicker that can weaken the effectiveness of the training

Clicker training is spreading worldwide as people discover and practice it, learning from each other and from teaching aids like this book. Nevertheless it is new to many people who participate in dog shows. The very concept of a training system based on science can seem unnatural in a world so deeply rooted in tradition. Some handlers and breeders may resent new methods and say so. Some may pick up the clicker but have no clear idea of its purpose, and therefore get mixed or poor results. Some might jump to false conclusions about what the clicker does or doesn't do, and then pass on their misconceptions to others. Here are some common misconceptions about the clicker and its uses, together with suggestions for more effective alternatives.

The clicker is not a way to get your dog's attention.

If you use it like some sort of squeaky toy, to capture the dog's interest, it will lose its virtues as an information system and quickly become just one more silly noise to be ignored.

The click is not an inhuman mechanical device that replaces your voice and praise.

You can go on using praise all you want; just click first. The click sound thus identifies for the dog exactly what all your affection is in honor of. You might say that the voice is analogue—like a picture, it carries a lot of information, including your mood, your energy level, and so on—whereas the click is digital: it means one thing only: Yes. Dogs learn in an instant that they can make you click, and how to do it again. It's clear communication, and that rapidly leads to a stronger bond between you, one of the many delightful side effects of this kind of training.

The clicker is not a permanent necessity.

The clicker conveys information, especially new information, far more accurately than words can. It's invaluable for teaching new behavior. However, once the behavior has been learned, and you have established a cue that tells the animal when to do it, you can replace the mechanical click with a mouth sound, a word, or even a smile or a nod, to let the animal know it did the right thing and has earned a treat.

You do not have to use a clicker in public or on the show grounds.

There are no rules against noisemakers and treats in the conformation show ring; handlers use all kinds of attention-getting devices. Because the clicker is new, however, and because not everyone understands it, other exhibitors sometimes object to this particular sound. There's an easy and polite solution. Use the clicker at home to teach the dog the right way to stack or gait or pose. Choose and establish a substitute marker signal. When you are actually on the show grounds, leave your clicker in your pocket and use your substitute signal to say "That's right!" You'll attract less attention from bystanders, and your dog will understand you just the same.

The click is not a way to call your dog and make it come.

Yes, it will work for that, at least at first; the dog hears the click and rushes over to get its treat. Many novice clickers whose dogs are inveterate off-leash disobeyers find that impressive. But the effect will fade fast. The click always means, "You're doing the right thing, and I'm going to pay you for it." If you call the dog, and it doesn't come, and you click anyway, you're saying "Stay over there and ignore me, and I'll pay you for it." Do it more than two or three times and you will actually be training the dog to stay away in order to get called and get a treat.

The clicker is not a threat or a warning of punishment.

Oh, you *could* use it for that, instead of for good news, but why would you want to? Anyone can threaten. It takes brains to get results without punishing—and you get better results.

The click is not just for puppies, or for pets, or for tricks.

Clicker training works for any dog, any breed, at any age, and for any purpose. It's good for "dumb" dogs that simply quit if handled forcefully (how dumb is that?) Clicker training is especially wonderful for all those breeds that are considered "hard to train." It engages zestful participation from independent terriers. It causes single-minded hounds to focus on something besides scent. It's perfect for toy dogs, who are often incredibly quick learners.

Clicker training is not a fad or the invention or method of some particular guru (including me).

Clicker training is a technology, based on behavioral science. This technology is rapidly overtaking traditional methods in the handling and management of all kinds of animals, including dogs, horses, birds, cats, and all the once untouchable species of wild birds and animals in zoos. There are human applications, too, such as physical therapy and sports training.

Clicker training is not obscure or hard to learn; it's just different.

This system doesn't depend on dominance or "respect," like traditional animal training. It's a mutual agreement based on accurate communication; it works without causing fear or requiring submissiveness. Because it's so different from traditional training and handling, children and complete novices often have an easier time learning clicker training than highly experienced handlers do. The best way to learn it is just to pick up a clicker and treats and get started. The dog will teach you.

CHAPTER 16

THE CLICKER LITTER:

Early clicker training will improve your puppies' chances of getting along in the world

How soon can you begin training puppies? As soon as their eyes and ears are open, according to some breeders, who are using the clicker on whole litters of pups, even before they are weaned. Why would you want to do that? Well, the clicker means good things are coming. The puppy that makes that connection can then learn that its own actions sometimes *cause* those clicks that lead to treats. And the puppy *that* makes that discovery has a big start on a happy future.

Here's how it works. As soon as supplemental feeding begins, the litter owner clicks as the pan of food is set down among the puppies. Some people click just once, and some click as each puppy nose actually reaches the food. Police officer Steve White, who breeds German shepherds, begins clicking even earlier, every time the dam goes into the litter box to nurse her babies—surely a very important event for the pups.

After some exposure to the clicker, begin taking each puppy away from the litter for a short session on its own. Click and treat. A dab of pureed baby food meat on the tip of your finger makes a great treat, even for the tiniest breed. Then pick something the puppy happens to do, such as lifting a front paw, and click as the paw goes up. It may take ten or more clicks before the puppy begins lifting the paw on purpose; but then you'll be amazed at how enthusiastic the puppy becomes. "Hey, look! I can make that huge person give me food, just by doing *this*!"

...these brief lessons can transform a puppy of five weeks or so from an oblivious blob into an eager, observant learner.

Choose any simple behavior at first: it doesn't need to be something useful. A sit, spin, wave, play bow, back up, or lie down are all possibilities. You can teach all the puppies the same behavior or, if you have them identified individually, teach them each something different. Don't try to coax or lure your students into a particular behavior; you want each puppy to discover that its own actions make you click. This teaches the puppy a major life lesson: "I like to find out what people want me to do." That discovery won't happen if the puppy learns to just wait to be shown what to do.

How much time does this take, in your busy life? Two or three clicker lessons, of no more than two to five minutes each, are enough to develop some cute little behavior. No need for a lot of drilling; once a puppy learns what to do for a click, it won't forget.

More importantly, these brief lessons can transform a puppy of five weeks or so from an oblivious blob into an eager, observant learner.

You can capitalize on this awakened state in many ways. For example, when people come up to the litter box, do the puppies rush over and leap on the walls, begging for attention? Probably. So use clicker sense and make a new rule—a rule for puppies and for family and for visitors, too: only puppies that are *sitting* get petted, or lifted out of the pen. It doesn't take long to get the whole litter sitting; and you can click them all at once, for doing that. Now, when supper comes, the puppies will have to sit and be clicked before the dish goes down. Instead of repeatedly and unintentionally reinforcing jumping up, a behavior most dog owners really hate, you are building a bunch of pups with better manners than that, even before they leave for their new homes.

"Come when called" is another skill the whole litter can learn with clicks and treats, and a fun one for children to teach. Two or three children can take turns calling a puppy back and forth between them, clicking and treating when the puppy goes to the child who called. If you're a breeder, you're going to give your buyers a puppy that already has a head start on this important behavior.

How far can you go? Training with absolutely no corrections, just informative clicks and enjoyable treats, you can go a long way, even with a baby. When my last Border terrier puppy, a long-distance purchase bought sight unseen, arrived on the airplane, she was just nine weeks old. I brought her home, set her down, and gave her a little toy. She picked it up, carried it to me, and dropped it at my feet. Surely this is an accident? I thought. I tossed it. She went and got it, brought it back, and dropped it at my feet again. Using clicks and treats, the breeder, as a treat for *me*, had taught this tiny puppy a nice retrieve! Furthermore I never had to teach her anything more about the retrieve; she was a reliable and dedicated retriever of toys, balls, and other small objects, for the rest of her life.

Breeders with clicker-trained litters usually give their buyers a demonstration of what the puppy has learned, and a beginner's clicker training kit, or kit and video, to take home (available at www.clickertraining.com.) People love taking home a puppy that already knows a trick; what a smart dog! Going home with a clicker and instruction book encourages them to try clicker training themselves, right away, when they are most motivated. And your early work starts them off with an attentive and cooperative pup that is ready to learn more—a puppy that has a far better chance of fitting in to its new world than a puppy starting from zero.

Melinda Johnson, a long-time breeder of soft-coated Wheaten terriers, began clicker training litters several years ago. Like many breeders, Melinda will always take a dog back if it doesn't work out in its new home. Melinda reports that since she started clicking litters, her return rate has dropped to zero; and her file of letters from thrilled and happy owners has grown enormously.

"Smartest, most attentive dog I've ever had."
"A laugh a minute, how did we get along without her?"

These puppies still have a lot to learn, of course. But a lot of potential troublesome behavior has been averted before it even develops. And the pups start their new lives knowing how to learn, and ready and eager to learn more.

APPENDICES

Appendix A:

Fifteen tips for getting started with the clicker

Appendix B:

The ten laws of shaping

Appendix C:

About target training

Appendix D:

The eight ways of getting rid of behavior you don't want.

APPENDIX A

FIFTEEN TIPS FOR GETTING STARTED WITH THE CLICKER

Clicker training is a new, science-based way to communicate with your pet. It's easier to learn than standard command-based training. You can clicker train any kind of animal, of any age. Puppies love it. Old dogs learn new tricks. You can clicker-train cats, birds, and other pets as well. Here are some simple tips to get you started.

1. **Push and release the springy end of the clicker, making a two-toned click.** Then treat. Keep the treats small. Use a delicious treat at first: for a dog or cat, little cubes of roast chicken, not a lump of kibble.

2. **Click during the desired behavior, not after it is completed.** The timing of the click is crucial. Don't be dismayed if your pet stops the behavior when it hears the click. The click ends the behavior. Give the treat after that; the exact timing of the treat is not important.

3. **Click when your dog or other pet does something you like.** Begin with something easy that the pet is likely to do on its own. (Ideas: sit; come toward you; touch your hand with its nose; lift a foot; touch and follow a target object such as a pencil or a spoon.)

4. **Click only once (in-out.)** If you want to express special enthusiasm, increase the number of treats, not the number of clicks.

5. **Keep practice sessions short.** Much more is learned in three sessions of five minutes each than in an hour of boring repetition. You can get dramatic results, and teach your pet many new things, by fitting in a few clicks a day here and there in your normal routine.

6. **Fix bad behavior by clicking good behavior.** Click the puppy for relieving itself in the proper spot. Click for paws on the ground, not on the visitors. Instead of scolding for making noise, click for silence. Cure leash-pulling by clicking and treating those moments when the leash happens to go slack.

7. **Click for voluntary (or accidental) movements toward your goal.** You may coax or lure the animal into a movement or position, but don't push, pull, or hold it. Let the animal discover how to do the behavior on its own. If you need a leash for safety's sake, loop it over your shoulder or tie it to your belt.

8. **Don't wait for the "whole picture" or the perfect behavior.** Click and treat for small movements in the right direction. You want the dog to sit, and it starts to crouch in back: click. You want it to come when called, and it takes a few steps your way: click.

9. **Keep raising your goal.** As soon as you have a good response-when a dog, for example, is voluntarily lying down, coming toward you, or sitting repeatedly, start asking for more. Wait a few beats, until the dog stays down a little longer, comes a little further, sits a little faster. Then click. This is called "shaping" a behavior.

10. **When your animal has learned to do something for clicks, it will begin showing you the behavior spontaneously, trying to get you to click.** Now is the time to begin offering a cue, such as a word or a hand signal. Start clicking for that behavior if it happens during or after the cue and ignoring it when the cue wasn't given.

11. **Don't order the animal around; clicker training is not command-based.** If your pet does not respond to a cue, it is not disobeying; it just hasn't learned the cue completely. Find more ways to cue it and click it for the desired behavior. Try working in a quieter, less distracting place for a while. If you have more than one pet, separate them for training, and let them take turns.

12 **Carry a clicker with you to "catch" cute behaviors like cocking the head, chasing the tail, or holding up one foot.** You can click for many different behaviors, whenever you happen to notice them, without confusing your pet.

13 **If you get mad, put the clicker away.** Don't mix scoldings, leash-jerking, and correction training with clicker training; you will lose the animal's confidence in the clicker and perhaps in you.

14 **If you are not making progress with a particular behavior, you are probably clicking too late.** Accurate timing is important. Get someone else to watch you, and perhaps to click for you, a few times.

15 **Above all, have fun.** Clicker-training is a wonderful way to enrich your relationship with any learner.

APPENDIX B

THE TEN LAWS OF SHAPING

Adapted from Chapter 2 of Don't Shoot the Dog *revised edition by Karen Pryor*

1. Raise criteria in increments small enough so that the subject always has a realistic chance of reinforcement.

2. Train one aspect of any particular behavior at a time. Don't try to shape for two criteria (two aspects of the behavior) simultaneously.

3. During shaping, put the current level of response on a variable ratio schedule of reinforcement before adding or raising the criteria.

4. When introducing a new criterion or aspect of the behavioral skill, temporarily relax the old ones.

5. Stay ahead of your subject: Plan your shaping program completely so that if the subject makes sudden progress, you are aware of what to reinforce next.

6. Don't change trainers in midstream. You can have several trainers per trainee, but stick to one shaper per behavior.

7. If one shaping procedure is not eliciting progress, find another. There are as many ways to get behavior as there are trainers to think them up.

8. Don't interrupt a training session gratuitously; that constitutes a punishment.

9. If behavior deteriorates, "Go back to kindergarten." Quickly review the whole shaping process with a series of easily earned reinforcers.

10. End each session on a high note if possible, but in any case quit while you're ahead.

APPENDIX C

ABOUT TARGET TRAINING

Targets can be a versatile training aid. Dogs, cats, and other animals easily learn to touch a target for a click and a treat. Touching a target is also an easy behavior for new clicker trainers to teach.

You can use any object for a target. A pencil or the end of your finger or the dot of light from a laser pointer all make good targets. Agility trainers use a plastic food container lid placed on the ground or obstacle wherever they want the dog to stop. Post-It notes make good movable targets. Horse clicker trainers have taken to using orange plastic traffic cones that horses can easily see at a distance.

Many dog trainers use folding aluminum target sticks (available from www.clickertraining.com) which can be carried conveniently in a pocket or bag and unfolded when needed. The dog is taught to touch and then to follow the tip of the stick. When the dog will freely follow the stick, you can lead the dog wherever you want it to go, instead of having to pull, push, lift, or coax it. You can target the dog into the car, onto a grooming table, over jumps, or into the correct position for gaiting or stacking. You can put the target upright in the ground to teach the dog to go away from you. With the target stick you can teach many tricks and useful skills, such as closing a door, turning a light switch on and off, and retrieving objects by name.

Here are some basic tips for developing the behavior of targeting. For detailed instruction on using the target stick in many kinds of advanced training see Morgan Spector's very complete guide, *Clicker Training for Obedience* (see Resources.)

Getting started with target training

1. Rub some food on the tip of the target stick and encourage the dog to sniff it. Click for looking at the stick, for nosing it, licking it, and bumping it. Give the dog a treat after each click. Repeat several times, putting the end of the stick an inch or two from the dog's nose each time.

2. Move the stick so its end is above the dog's nose, below it, to the left, and to the right, clicking for touches in each direction. Move it away a little, and click if the dog takes a step toward it. Try walking with the dog and the stick; sometimes the dog will catch on faster if it gets clicked while moving.

3. See if you can get the dog to stand on its hind legs to reach the tip of the stick, or bow down to reach to the floor. Settle for small movements at first; make it easy for the dog, not hard.

4. Keep your sessions short; three or four minutes at a time is plenty. Keep the stick and some treats handy, perhaps in the kitchen, so you can do a little target training several times a day. Some dogs will catch on in a single session, and begin racing for a chance to touch the stick; others may take five or six sessions just to learn to touch it with confidence.

5. Watch for signs of understanding: a wagging tail is a good sign. When the dog is eagerly touching and following the stick, and perhaps grabbing at it when you aren't even asking for that, raise your criteria. Start asking him to touch it two or three times for a single click and treat, or to follow it for several steps.

6. Omit the click if the dog mouths or bites the target stick, or touches it along the side rather than at the end. Your dog will not mind the omitted clicks, but will try harder to find out exactly what he needs to do to get you to click him again.

7. Now you can use the target stick to teach other behavior. If you are interested in agility training, you can use the target stick to teach the dog the obstacles, and to indicate contact points. An obedience trainer could use the target to teach go-outs and the drop on recall.

8. Use the target to teach the dog to walk beside you on a loose leash, or out in front of you in "parade" position for the conformation show ring.

9. You can transfer the behavior to other targets. Yellow sticky notes can be used as targets on furniture or on/off switches, or to teach the dog to retrieve specific items such as the TV remote. The red dot of a laser pointer can be a wonderful target for working your dog at a distance; a laser pointer can be useful in tracking and other scent work, in agility, and in police work.

10. Above all, have fun with target training and enjoy this new way of communicating with your dog.

APPENDIX D

THE EIGHT WAYS OF GETTING RID OF BEHAVIOR YOU DON'T WANT

Anything you do to get rid of behavior you don't want will fall into one of the following eight methods. The first four are the 'bad fairies,' the methods that have neither kindness nor special efficacy to recommend them. The second four are the 'good fairies,' the approaches that involve positive reinforcement and some understanding of behavior, and that are highly likely to work. This material, specially adapted for the show dog owner, is based on Chapter 4 of *Don't Shoot the Dog!* by Karen Pryor.

Method 1. Shoot the animal.

This definitely works. Get rid of the animal, by whatever means, and you will never have to deal with that particular behavior from that particular subject again. Method One is a common solution, in the dog show world, to a dog that "won't show." Give the dog away and buy a new and more expensive dog.

Method 2. Punishment.

Everybody's favorite, in spite of the fact that it almost never really works. In the show ring leash jerks are the commonest punishment (euphemistically called corrections) but I have also seen dogs stepped on, yanked off their feet, kneed in the ribs, and ear-pinched for not paying attention, for failure to obey a command, and for misbehavior such as growling at the handler. Punishment does not improve a show dog's attitude.

Method 3. Negative reinforcement.

This does not mean doing something negative to the dog when it makes a mistake: it means *removing* something negative when the dog does something right. For example, during gaiting and stacking in the show ring many handlers hold the leash high over the dog's head, dragging the dog upward. An appropriate use of negative reinforcement would be to slacken the leash whenever the dog holds its head high voluntarily.

Method 4. Extinction.

Letting the behavior go away by itself. For example, playfulness in a puppy, or overexcitement in any dog making its first trip into the show ring, will go away with or without training as the dog matures and becomes accustomed to the show environment. Clicking for calmness, or clicker training specific alternative behaviors such as focusing on a target, can speed the process of desensitization and help extinguish overreactions.

Method 5. Train an incompatible behavior.

The dog sniffs the ground all the time in the ring? Click it for keeping its nose in the air for two steps, then three, then five, then a ring length, then for longer and longer distances. The dog is being paid to keep its head up; it cannot do that and sniff the ground at the same time. Eventually just putting on a show collar and lead can become a cue for "Keep your nose off the ground." Training an incompatible behavior—and paying for it, with the treats one is allowed to carry in the show ring—is much more effective than either punishing the sniff (which encourages the dog to try to sneak in sniffs when you're not looking) or the physical intervention of hauling the dog's head into the air by the leash, which will give you a sore arm by the end of the day.

Method 6: Put the behavior on cue. (Then you almost never give the cue.)

This is an elegant way of getting rid of unwanted behavior, but so counter-intuitive that most people just can't bring themselves to try it. Click the behavior until the dog is offering it for the click. Add a cue.

Reinforce the behavior when you have cued it, ignore it when you haven't. The behavior will disappear except when cued. This is one way to get rid of puppyish appeasement behaviors such as frantic face-licking; pawing and begging; jumping up; intrusive sniffing; barking and whining; scratching at doors; and (trust me) submissive urination. These are all innate puppy-to-adult-dog behaviors that we often intensify both by getting angry (causing appeasement behavior to increase) and by inadvertent reinforcement.

Method 7. Shape the absence of the behavior.

Reinforce everything that is not the undesired behavior. This method is particularly appropriate with fearful or aggressive dogs. If the dog does *anything* normal, click. And treat. Keep the sessions short, keep the reinforcements coming thick and strong, once every ten seconds at least, and repeat as desired.

Method 8. Change the motivation.

Example: the dog in the yard that barks all night, disturbing the neighbors. This is a lonely and frightened dog. Let the dog sleep in the house. Problem solved. In this particular instance, identifying the cause of the behavior and eliminating it is a useful approach. Too often, however, modern dog owners try to solve any and all behavior problems by analyzing or explaining the dog's 'reasons' for acting that way ("It's territorial aggression!" or "He's being dominant!"). Except for genetic or medical problems, speculating as to why a behavior is happening is often of little practical benefit. Instead, identify the behavior specifically, then use one of the other eight methods to replace that behavior with something more acceptable.

RESOURCES

Showing your own dog in the breed ring

Clicker training is one skill; showing your own dog is another. Excellent books and videos are available covering show rules, show tactics, grooming, the special skills of professional handlers, breeding your own show dogs, and other aspects of the sport. None of these books (so far) are clicker-based; but it's important to realize that anything that can be trained, can be trained with the clicker. You can draw freely on the expertise of these experienced authors; just substitute your own clicker plans for any instructions that involve choke collars, discipline, 'corrections,' or other aversive methods. Clicker training works just as well, and without risking the dog's good will and enthusiasm for showing. A good online and mailorder source for breed and show ring books: ***www.dogwise.com***

The science underlying clicker training

The most popular book on positive reinforcement and operant conditioning, and how they work with people and animals both, is ***Don't Shoot the Dog! The new art and science of training***, by Karen Pryor, Bantam Books. Get the revised edition, published in 1999; earlier versions are out of print. ***Don't Shoot the Dog*** by Karen Pryor, $12.95 plus S&H through bookstores, Amazon, or www.clickertraining.com

Clicker training books for dog owners

Getting Started: Clicker Training for Dogs Kit with book by Karen Pryor, treats, two clickers, pocket instructions, $16.95 plus S&H from www.clickertraining.com. or in PetCo stores nationwide.

Clicking with Your Dog, Step-by-Step in Pictures, by Peggy Tillman; Sunshine Books. 100 behaviors, illustrated click by click in simple drawings; lots of good management and care information for the beginning or inexperienced dog owner as well. Very popular and well reviewed. $29.95 plus S&H through bookstores, Amazon, or www.clickertraining.com.

Clicker Training for Obedience by Morgan Spector; Sunshine Books. By far the most detailed clicker training dog book on the market. Even if you are not an obedience competitor, this book is a rich source of clicker theory, illustrations, and practice, useful in any kind of advanced training application $29.95 plus S&H through bookstores, Amazon, or www.clickertraining.com.

Getting Started: Clicker Training for Cats, by Karen Pryor. This delightful book opens up a whole new way to enjoy your cat, and to keep your cat active, happy, amused, and interested in life and in you. Well reviewed and very popular with people who began clicker training with their dogs—and wondered about the cat. Book alone $12.95. In a Kit with two clickers, treats, pocket instructions, $16.95 from www.clickertraining.com.

Clicker Training for Your Horse, by Alexandra Kurland. The 'bible' on clicking with horses. Revolutionary, and wonderful, producing horses that are gentle, friendly, willing, safe, and fun to have around. $29.95 from www.clicker-training.com.

Getting Started: Clicker Training for Horses Kit, by Alexandra Kurland. Book, two clickers, treats, pocket instructions $16.95 from www.clickertraining.com.

Videos

Clicker Magic

With Karen Pryor. 20 real life clicker training sessions with a variety of trainers and animals, including puppies, senior dogs, a mule, a cat, and even a fish. A great introduction to what clicker training can do. 55 minutes VHS. Comes with a free clicker. $39.95 from www.clickertraining.com.

Puppy Love: Raise Your Dog the Clicker Way

With Karen Pryor and Carolyn Clark

An introduction to puppy-raising and the clicker. Includes housetraining, crating, socialization, tricks, grooming and veterinary care. $24.95 from www.clickertraining.com

Equipment

Clickers

Three-pack with instructions $6.95
Five-pack with instructions $10.00
Thirty-pack $37.50
Clip-on clicker for wrist or belt $7.95
Extendable clicker $7.95

Target Stick

Folding aluminum target stick $16.95

Prices valid as of December 2001 and are subject to change without notice.

Online resources

Karen Pryor's Web site, ***www.clickertraining.com***, contains a wealth of articles and background information on clicker training, a clicker trained honor roll, clicker store, interactive clicker community, links to clicker lists and Web sites, and information on locating a clicker trainer in your area.

Acknowledgments

I owe many thanks to Beth Edelman, Josh White, and Alan Gomberg, the editors I have been privileged to work with at the *AKC Gazette*. I am also most grateful to AKC publisher George Berger, for his enthusiasm and support for this project. Special thanks go to Paige Elliott for her valued friendship and her sound counsel on how dogs should be shown. I also want to thank my class participants, scientific colleagues, fellow dog show enthusiasts, and the host of clicker trainers whose insights, experiences, and clever clicker dogs contributed to my understanding of this technology and its applications. Any errors or oversights in this book are not theirs, but mine alone.

Karen Pryor, January 1, 2002

About the author

Karen Pryor is a behavioral biologist and author whose popular books range over a wide variety of subjects, from mothers and infants to dolphin social systems. As a founder and curator of Sea Life Park in Hawaii she pioneered many of the techniques that are now standard worldwide in the training of marine mammals. She is the author of the leading textbook on positive reinforcement in humans and animals, *Don't Shoot the Dog! The New Art of Teaching and Training.* As CEO of Sunshine Books, Inc., and its affiliate Clicker Training.com (www.clickertraining.com) she has encouraged the development of many new uses for reinforcement-based teaching, including horse training, behavioral management of cats, birds, and zoo animals, shelter management, veterinary care, and the training of human sports and skills. She has three children and seven grandchildren, and lives in Boston with a border terrier, a harlequin poodle, and a Burmese cat, all of whom are clicker trained.

Photo Credits

©AKC, photo by Chet Jezierski, front cover; ©AKC, p. 1; ©AKC, photo by Mary Bloom, pp. v (center and bottom), 3, 14, 20, 25, 40–41; ©AKC, photo by DogPhoto.com, p. 29; ©AKC, photo by Robert Sternlieb, pp. vi (top), 7, 9, 35; ©Mary Bloom, pp. x, 5; ©Kent and Donna Dannen, pp. v (top), xiii, 18, 23; ©DogPhoto.com, pp. vi (third from top), 59; ©Espen Engh, p. 32; ©Karen Pryor, pp. vi (second from top, bottom), 15, 44, 45, 46, 49, 52, 53, 56, 58, 63, 71, 73, back cover; ©Sea Life Park, p. viii; ©Susan Smith, p. xi